Social Perspectives on Ancient Lives from Paleoethnobotanical Data

Matthew P. Sayre · Maria C. Bruno
Editors

Social Perspectives on Ancient Lives from Paleoethnobotanical Data

Springer

Editors
Matthew P. Sayre
Department of Anthropology
University of South Dakota
Vermillion, SD
USA

Maria C. Bruno
Department of Anthropology
and Archaeology
Dickinson College
Carlisle, PA
USA

ISBN 978-3-319-85000-9 ISBN 978-3-319-52849-6 (eBook)
DOI 10.1007/978-3-319-52849-6

Printed on acid-free paper

This Springer imprint is published by Springer Nature
The registered company is Springer International Publishing AG
The registered company address is: Gewerbestrasse 11, 6330 Cham, Switzerland

The original version of the book was revised:
The book title was corrected from "Social Perspectives on Ancient Lives from Paleothnobotanical Data" to "Social Perspectives on Ancient Lives from Paleoethnobotanical Data".

Contents

Editors and Contributors

About the Editors

Dr. Matthew P. Sayre is currently Assistant Professor at the University of South Dakota. He teaches courses in Archaeology, Biological Anthropology, and Historical Ecology. He is also active in the development and teaching of the new Sustainability Major. His research focuses on the past ecology, ritual, and production practices of people in the Andean region of South America. He has worked at the UNESCO World Heritage site of Chavin de Huantar, Peru over the last decade and leads a field project focused on the domestic settlements in the La Banda sector of the site.

Dr. Maria C. Bruno is currently Assistant Professor at Dickinson College where she teaches courses in Archaeology and World Prehistory, Environmental Archaeology, and Archaeological Method and Theory. She is also actively engaged in Dickinson College's interdisciplinary Center for Sustainability Education. Her research focuses on the environmental, technological, social, and political dynamics of ancient agricultural systems in South America, particularly in the Lake Titicaca Basin of the Andes and in the Llanos de Moxos region of the Amazon Basin.

Contributors

Dominique Bosquet SPW-DGO4-Service de l'Archéologie en province de Brabant Wallon, Wavre, Belgium

Maria C. Bruno Department of Anthropology and Archaeology, Dickinson College, Carlisle, USA

Alexandre Chevalier Royal Belgian Institute of Natural Sciences, Brussels, Belgium

Katherine L. Chiou Department of Anthropology, University of California, Berkeley, USA

Margaret W. Conkey Department of Anthropology, University of California, Berkeley, USA

Rob Q. Cuthrell Archaeological Research Facility, University of California, Berkeley, USA

Alan Farahani Cotsen Institute of Archaeology, University of California, Los Angeles, USA

Gayle J. Fritz Department of Anthropology, Washington University, St. Louis, USA

Anna Harkey Department of Sociology and Anthropology, Wofford College, Spartanburg, USA

Christine A. Hastorf Department of Anthropology and Archaeological Research Facility, University of California, Berkeley, USA

Logan Kistler Department of Anthropology, National Museum of Natural History, Smithsonian Institution, Washington, D.C., USA

M.A. Korstanje Instituto de Arqueología y Museo (FCN e IML, Universidad Nacional de Tucumán), Instituto Superior de Estudios Sociales (CONICET/UNT), San Miguel de Tucumán, Argentina

BrieAnna S. Langlie Department of Anthropology, Loyola University Chicago, Chicago, USA

Christopher T. Morehart Department of Anthropology, Arizona State University, Tempe, AZ, USA

Shanti Morell-Hart Department of Anthropology, McMaster University, Hamilton, Canada

Deborah M. Pearsall Department of Anthropology, University of Missouri, Columbia, MO, USA

Matthew P. Sayre Department of Anthropology, University of South Dakota, Vermillion, SD, USA

Payson D. Sheets Department of Anthropology, University of Colorado, Boulder, USA

Bruce D. Smith Program in Human Ecology and Archaeobiology and Department of Anthropology, National Museum of Natural History, Smithsonian Institution, Washington, D.C., USA

William T. Whitehead SWCA Environmental Consultants, Albuquerque, NM, USA

Chapter 1
Social Paleoethnobotany: New Contributions to Archaeological Theory and Practice

Maria C. Bruno and Matthew P. Sayre

Introduction

Paleoethnobotany or archaeobotany,[1] simply defined as the study of plant remains from archaeological sites, has become a central component of archaeological practice across the globe. Not only are its methods for the recovery and analysis of a wide range of plant remains regular elements of the most rigorous academic and contract archaeological projects today, but research agendas are increasingly informed by questions that can be answered with archaeological plant remains (Marston et al. 2015: 9–10). Although still sometimes viewed as merely a methodological specialization, paleoethnobotanists have a long history of using the data they produce to address larger questions about the human past and contribute to broader theoretical discussions in the field of archaeology (see Hastorf 1999; Marston et al. 2015; Pearsall 2015; VanDerwarker et al. 2016 for reviews of the discipline). Paleoethnobotany has contributed rather substantially to theories of the nature of human–environmental interactions and subsistence change, particularly the origins of agriculture. It has also, but in a more limited way, contributed to theories about social and political processes, especially with regard to food production and consumption among complex societies. This volume adds to the growing arena of social paleoethnobotany (Morehart and Morell–Hart 2013) with a series of papers exploring dynamic aspects of past social life, particularly the

[1]The term paleoethnobotany is more common in North America whereas archaeobotany is the dominant term in Europe. In this volume we will refer to the field as paleoethnobotany.

M.C. Bruno (✉)
Department of Anthropology and Archaeology, Dickinson College, Carlisle, USA
e-mail: brunom@dickinson.edu

M.P. Sayre
Department of Anthropology, University of South Dakota, Vermillion, USA
e-mail: Matthew.Sayre@usd.edu

M.P. Sayre and M.C. Bruno (eds.), *Social Perspectives on Ancient Lives from Paleoethnobotanical Data*, DOI 10.1007/978-3-319-52849-6_1

day-to-day practices and politics of procuring, preparing, and consuming plants across range of places, times, and contexts. While many of the papers bring social perspectives to the more traditional realms of paleoethnobotanical research such as defining subsistence practices and domestic activities, they also delve into topics that are still emerging within the sub-field including ritual and gender. Together, these papers shed light on ways in which the specialized analysis of plant remains can contribute to theory building and advancing archaeological understanding of past lifeways.

Although not the focus of this volume, methodological advancements in paleoethnobotany have been central to its growing importance within archaeology as well (Hastorf and Popper 1989; Marston et al. 2015; Pearsall 2015). From the advent of water flotation for the recovery of macrobotanical remains, to the chemical extractions of phytoliths and starch grains, and now technologies to collect *a*DNA from archaeological plant remains, a powerful suite of botanical techniques are growing our paleoethnobotanical databases. The ideas presented in these chapters are founded on the rigorous application of a wide range of paleoethnobotanical methods. As Pearsall (this volume) highlights in her final chapter, the use of multiple lines of paleoethnobotanical evidence is essential to creating robust explanations regarding past human activities.

Archaeological Theory and Paleoethnobotany

The development of paleoethnobotany has followed some of the general trends apparent in the broader field of archaeology in North America and Europe (Hastorf and Johannessen 1996; Morell–Hart 2015), although its trajectory is also linked to trends in the economics and politics of archaeology as a profession (Morehart and Morell–Hart 2013). The earliest studies of plant remains from archaeological sites were the result of fortuitous discoveries of plants preserved in unique contexts such as Egyptian tombs, water-logged sites of Europe, and in extremely dry sites in the Andes (Ford 1979; Renfrew 1973). In these early cases, botanists were sought to identify the species recovered and their findings presented as an appendix at the end of a report or book (Reitz et al. 2008). As with other material culture, the plants encountered at these sites entered into descriptive summaries of past culture histories contributing to growing knowledge about past foodways, construction, and crafting materials.

As archaeologists began to move beyond description to address broader questions about the human experience, the formal sub-field of paleoethnobotany solidified. Practitioners not only developed more precise and systematic methods for the recovery and identification of a wide range of plant remains but they did so in order to address the complex interrelationships of plants and humans in the past (Ford 1994; Hastorf and Popper 1989). Although the early advancement of paleoethnobotany certainly coincided with the New Archaeology and "processual" approaches with its emphasis on environmental adaptations and subsistence, its roots in ethnobotany also made relevant the social and political nature of human–plant interactions (Ford 1994; Hastorf 1990; Jones 1941; Lentz 1991). Paleoethnobotanists were thus among the first of the environmental archaeologists to evaluate issues such as meaning and gender using their datasets within the "post-processual" movement (Fritz 1990; Hastorf 1991, 1993; Hastorf and Johannessen 1996; Jones 1985; Watson and Kennedy 1991). In our analysis of theory and paleoethnobotany, we aim to break down the traditional "processual/post-processual" division and examine the unique ways in which paleoethnobotanists have used their datasets to contribute to a wide range of theoretical issues. While we do advocate for greater use of the social approaches in this volume, we recognize that paleoethnobotanists utilizing evolutionary/biological paradigms have made efforts to take into account the social and political nature of human–plant interactions (Gremillion 2015; Rosen 2007; Smith 2015).

To date, paleoethnobotanical contributions to archaeological theory have largely engaged with paradigms that can be categorized as evolutionary, ecological, and biological. Some of the most prominent North American paleoethnobotanists including Kristen Gremillion, Arlene Miller Rosen, Deborah Pearsall, Dolores Piperno, and Bruce Smith have written major treaties utilizing frameworks such as human behavioral ecology and niche construction theory (e.g., Piperno and Pearsall 1998; Rosen 2007; Smith 2011). It is noteworthy that these paleoethnobotanists have not only utilized these frameworks to analyze their data, but have substantially contributed to elaborating and modifying these theoretical approaches for archaeology (e.g., Gremillion 2002; Smith 2007). These approaches generally view humans and their interactions with plants as any other organisms on Earth, shaped by the processes of natural selection, resource availability, and feedback loops that create particular ecologies and evolutionary changes. Social and political aspects of human behavior are conceptualized as extensions of the biological adaptation of culture. Such approaches shed light on general, long-term processes such as significant shifts in subsistence associated with major climatic changes including the Pleistocene–Holocene transition and the ecological consequences of human interventions into plant communities such as resource depletion or disturbance, but often fail to capture the localized, small-scale social and/or political processes that contribute to or mitigate such changes. They are explicitly etic analyses of the past that do not necessarily aim to capture the lived experiences and cultural implications of these processes. Also, while there are several cases that seem to fit these models (see Gremillion 2015), there are often exceptions, and the exceptions are likely better explained by cultural and social preferences rather than "optimal," risk-reducing calculations (Chevalier and Boquet this volume; Ingold 1996).

While these evolutionary approaches remain dominant in the field, a growing number of paleoethnobotanists have argued for greater consideration of social factors shaping human–plant interactions and their roles in understanding the past human experience (e.g., Asouti and Fuller 2013; Bruno 2009; Denham 2005, 2009; Fairbairn 2008; Hastorf 1991, 1993; Hastorf and Johannessen 1996; Jones 1985; Morehart and Helmke 2008; Morehart and Morell–Hart 2013; Morell–Hart 2015). Plant remains are not only evidence of environmental phenomena but also of deeply cultural and political ones as well. Although humans interact with the plant world in a very material way, these interactions are imbued with meaning that shape various realms of human social life. In many human societies, plants are not simply viewed as resources to be used and calories to be consumed, but are central entities in the daily lives of people who collect, tend, cultivate, transform, destroy, and ingest them (e.g., Anderson 2000; Arnold 1996; Balée 1994; Descola 1994; Fowler 1996; Rival 1998). Many archaeologists are more interested in accessing these elements of the human past than testing evolutionary models.

Plant remains are particularly well-suited for querying the social and political dynamics of past human societies. Plant-related activities often occur on a daily basis, their remains regularly deposited, preserved (albeit not perfectly), and accessible to archaeologists who employ the wide range of recovery techniques now available. Plant-based craft and food production are carried out in prescribed ways resulting in culturally acceptable objects and meals that get reproduced over generations but are also loci for change and innovation (Atalay and Hastorf 2006; Bruno 2014; Morell–Hart 2015). Yet, not all plants are appropriate in all contexts. Plants are often used to mark particular times, people, and places, making important contributions to cultural definitions of contexts such as domestic, ritual, male, female, child, adult, life, and death (e.g., Allen 1988; Descola 1994; Hastorf 2003; Morehart et al. 2005; Sayre 2014). Thus, paleoethnobotanists are well positioned to engage with social theories of the human experience including agency, practice, identity (race, gender, class), and embodiment. Such approaches do not exclude environmental/ecological aspects of the human experience, however. Rather than viewing humans as simply adapting to environmental constraints, more social environmental paradigms consider the dialectic impact humans have on shaping local ecologies (and vice versa), as well as the creation of landscapes, the physical and conceptual entities wrought from human–environmental interactions (Balée and Erickson 2006; Crumley 1994; Ingold 2000).

An early example of a social paleoethnobotany comes from the pioneering studies of wood use in the Upper Mantaro Valley of the Peruvian Andes by Christine Hastorf and Sissel Johannessen (Hastorf and Johannessen 1996; Johannessen and Hastorf 1990). Paleoethnobotanists and paleoecologists have long used wood charcoal to identify species' composition in past environments, and the impact of human activities particularly burning and deforestation (Asouti and Austin 2005; Delcourt and Delcourt 1988; Smart and Hoffman 1988). Common economic and evolutionary models of human–tree interactions viewed wood simply as a resource to be exploited for fuel and construction, and its overuse would be inevitable through expanded agriculture and increasing human populations. Sites in the Upper Mantaro Valley contained evidence of pre-Hispanic life that span nearly 1000 years from approximately AD 500 to the time of Spanish contact AD 1532 (Hastorf and Johannessen 1991: 144). Hastorf and Johannessen found that wood charcoal was ubiquitous across all time periods slightly declining in the Middle Horizon periods (AD 900–1300) but increasing significantly in the Late Intermediate Periods (AD 1300–1460) and continuing into the Inka period. This increase in wood charcoal in the later periods, when agriculture intensified and populations grew, contradicted evolutionary and economic model expectations. Furthermore, not all tree species increased, but rather they found a shift from the predominance of *Polylepis* sp. and *Colletia* sp. in the earlier periods to an increase in *Buddlejia* sp. in the later periods.

It was clear to the authors that the pre-Hispanic populations were managing the tree populations. While this could have simply been for fuel, Hastorf and Johannessen examined the rich ethnographic and ethnohistoric record, and Hastorf conducted interviews to learn that trees were more than firewood, but entities with deep cultural and political significance in the region. Wood collecting was an important activity of daily life with both young and old individuals spending up to eight hours per days collecting fuel for their home fires. This not only kept the hearth burning but an abundant store of firewood "was a sign of prosperity" in Andean households (Johannessen and Hastorf 1990: 77). Given that this high, dry region is sparsely forested, there are written records of management practices that go back to the Inca, but based on the archaeological record even deeper. The Inka had explicit laws about where and when trees could be used, even designating sacred forests (Hastorf and Johannessen 1996: 75–76). Moreover, the scarcity and importance of certain species of wood gave them distinct meaning in terms of symbolic and political value.

Buddlejia sp., or *quishuar* in Quechua, was particularly significant to the Inka. There are historic memories of the Inka managing its cultivation and collecting it as tribute. Moreover, the *quishuar* (as well as other trees) were symbolic in terms of linking families to the land. A specific term—*mallqui*—was to refer to a cultivated tree (usually planted in a home compound). Trees being long-lived, connecting the earth, water, and sky, symbolized permanence on the landscape. Intriguingly, the term *mallqui* also meant ancestor mummies, which were often kept in places that represented a lineage's link to a particular territory (Hastorf and Johannessen 1996: 73–74).

Building from these ethnohistoric sources, Hastorf and Johannessen argued that the archaeological charcoal data were indicative of the antiquity of tree cultivation in the region beginning before the Inka with the Late Intermediate Period Wanka residents. They further argued that during this particularly contentious period of political history in the region trees, like mummies, may have been used to indicate territory. The management of trees thus served to maintain an important, consumable resource but also played a significant role in defining contested social and political spaces.

This example illustrates the social and political importance of plants, and demonstrates how patterns in plant data can force us to reconsider our assumptions about past human behavior. This can help us build more nuanced and robust theoretical models from which to think about broader trends in the past. With this example we contend that paleoethnobotanists are well positioned to make substantial contributions to social theories of human behavior and experience as they have to evolutionary and biological theories in the social sciences.

The diverse set of papers presented in this volume draw on the pioneering work of social paleoethnobotanists, such as Hastorf and Johannessen. Christine Hastorf's contributions, in particular, have not only been central to advancing paleoethnobotanical method and interpretation but making this specialized analysis an essential component of general archaeological practice and theory building. She was among the first paleoethnobotanists to use plant data to build social theories of the origins of agriculture (Hastorf 1998, 2006) and the emergence of the state (Hastorf 1993). Perhaps most significantly, she was central in shifting the view of plants as "lowly items" hardly worthy of study (Hastorf 1999: 57), to central units of archaeological research because of their social and political importance in the daily lives of people, particularly women (Hastorf 1991, 1998, 2001). In recognition of her work, she was the recipient of the Fryxell Award for Interdisciplinary Research at the 77th Annual Meeting of the Society for American Archaeology Meeting and several of the papers in this volume originated in the session organized in celebration of her award. Examples of her work are used as case studies throughout this volume.

New Studies in Social Paleoethnobotany

The first three chapters of the volume address "traditional" paleoethnobotanical issues related to subsistence: the collection of wild resources, the domestication of crops, and spread of agriculture. These chapters build on rich datasets of the spatial and temporal distribution of wild and domesticated plant species, and what these patterns indicate about the role these species played in the foodways of the regions examined. As discussed above, this has long been one of the realms where evolutionary theories predominate in the literature, and these chapters provide new insights into the more social processes that contributed to subsistence practices and change.

We begin with our only non-American contribution by Chevalier and Bosquet who examine the adoption of agriculture in Northwestern Europe through the analysis of five different botanical proxies from the early Neolithic (LBK) site of Remicourt "En Bia Flo II" in eastern Belgium. This in-depth study not only clarifies the timing of the entrance of domesticated species into the region but also interrogates the socio-political processes that shaped this profound transformation of local foodways. Chevalier and Bosquet compare the utility of Human Behavioral Ecological (HBE) and Historical Ecological (HE) paradigms for evaluating the adoption of agriculture in this region. They argue that of the two models, HE provides a more appropriate interpretive framework for it better accounts for the botanical evidence, which do not meet the strictly adaptationist expectations of an HBE model. For example, the forms of wheat and barley encountered in the sites are not those best adapted to the conditions in this region, but rather appear to be those preferred for cultural reasons, like flavor or cooking qualities. Furthermore, they show through pollen and wood analyses that the early farming populations did not simply adapt to local environmental conditions but modified the local flora to meet the needs of their farming systems and plant preferences.

Fritz, Bruno, Langlie, Smith, and Kistler bring together recent archaeobotanical and molecular data on the domestication and role of the important genus

Chenopodium in North and South America. By building on methods developed and modified in both regions, these authors show that it is not only possible to identify wild versus domesticated forms of these species but different varieties within them that represent the choices made by farmers based upon a mix of ecological, social, political, and cultural variables. Molecular studies demonstrate that *Chenopodium* was independently domesticated in eastern North America, Mesoamerica, and South America, revealing the individual agency of early gatherer–hunters throughout the Americas. While its ecological tendency to occupy human-disturbed spaces likely contributed to its domestication in both regions, comparison of the specific trajectories reveals distinct social and political histories of these chenopod species. Finally, in light of its resurgence in international markets today, this study reflects on how colonial incursions played a role in the "loss" of this genus's importance across the Americas.

These chapters are followed by Korstanje's interrogation of the common dichotomy of wild versus domestic, not in terms of differentiating wild progenitors from crops, but to understand the role of nondomesticated plant species in agricultural and pastoral societies. She does so by examining the well-preserved plant remains from rockshelters in Northwest Argentina. The consideration of rockshelters themselves challenges traditional archaeological categories of hunter–gatherer versus agriculturalist–pastoralist because such places have long been associated with the former but, in fact, provide rich data on plant use by the latter. Korstanje reveals the diversity of nondomesticated species that agricultural groups brought to and used in the rockshelters, and argues that although they were not planted and raised, the labor and social organization required to procure them was still significant, especially as many of them came from other regions.

The next three chapters also draw on a long tradition in paleoethnobotany of examining the spatial composition and distribution of plant remains, in conjunction with other archaeological remains, to elucidate the activities that took place in particular contexts (Hastorf 1988). Such studies have not only been central for clarifying the function of different types of contexts, but shed light on the particular practices that produced them and gave meaning to both daily life and special occasions.

The chapter by Farahani, Chiou, Cuthrell, Harkey, Morrell–Hart, Hastorf, and Sheets, while including paleoethnobotanical data, examine a range of other remarkably well-preserved objects that were left in place in a single household before the eruption of the Loma Caldera volcano at the site Joya de Cerén in El Salvador. Using Geographic Information Systems (GIS), the authors examine the spatial relationships of plant remains as well as ceramics, grinding stones, and animal remains, to shed light on the culinary practices typical of daily life at this community in seventh century AD. Their careful analysis of objects across various spaces within Household 1 provides a vivid rendition of the interrelated activities, or taskscapes, involved in daily foodways. The spatial association of several plant foods, including beans and corn near a duck that had been left behind, reminds us that domestic food stuffs are not always destined for humans but also provide food for other creatures that contribute to a household, including those that may eventually become food themselves.

Sayre and Whitehead examine how plant remains inform the interpretation of different spaces at the Peruvian Middle Horizon site of Conchopata, which was a provincial center of the Wari state between AD 650 and 1000. Archaeologists working at such sites often differentiate domestic versus ritual space using architecture and ceramic remains. Sayre and Whitehead argue that plant remains can provide new insight into the formal separation and designation of space as well as the activities carried out in domestic and ritual areas. Like Korstanje's paper that challenged the wild–domestic dichotomy, they explore the boundaries of the ritual–domestic divide and find areas where it is not so clear cut. They provide important insight into how the foods and drinks consumed in the ritual spaces are often prepared in the domestic ones, and that ritual can infuse daily activities as well. While there was some clear contexts of state-sponsored food and drink production at Conchopata, particularly for the maize beer, *chicha*, smaller spaces of ritual food and drink production are also hinted at in households.

Morehart also contributes to the anthropological study of ritual using paleoethnobotanical data to discern the temporality of particular rituals carried out at a shrine near Lake Xaltocan dating to the Epiclassic period (AD 600–900) in Mexico's northern Basin. Morehart draws upon the biological aspects of the plants encountered, including their lifecycles, edible part maturation season, particularly maize, and peak blooming period of flowers, to discern the seasonality of when rituals took place using specific plants. Then, he draws upon ethnographic and ethnohistoric information on the symbolism of these plants, especially as offerings to deities, in Mexican cosmology and the ceremonial calendar in order to elucidate the practices and meanings that contributed the ritual(s) conducted at this shrine. Through this analysis, Morehart argues that it is possible to access the immaterial, or meaning, of ritual in archaeological contexts through the study of plants.

The final two chapters of the volume examine the broader implications for a socially engaged paleoethnobotany in review of work by other paleoethnobotanists and those that contributed to this volume. Conkey places the work of paleoethnobotany in the broader history of feminist approaches to archaeology. She argues that the intersection of these two approaches reminds us that we do not just do theory or data, but that there is a relationship between these domains and they mutually inform one another. This perspective permits us to use botanical remains to inform not only our understanding of past subsistence but also on the nature of gender roles in societies. As the field has expanded, it has begun to investigate social questions that get to the heart of social identity. Pearsall provides summarizing remarks on the papers with special emphasis on how much paleoethnobotany has moved beyond diet and subsistence. She describes how the field has moved from a focus on methodology toward one that considers many of the broader implications of this research. This summary of the volume examines three major themes: seeking "the invisible" in the paleoethnobotanical record, revealing new aspects of ritual behavior, and using paleoethnobotany to gain new perspectives on the social realm. Her perspective is immensely valuable as she provides insights into how important it is to analyze botanical data from a variety of perspectives.

Conclusions

The contributors to this volume illustrate how one can bridge differences between the natural and social sciences through the more socially focused interpretations of botanical datasets. These papers demonstrate that paleoethnobotanists no longer simply provide a list of species for others to interpret but are integrating multiple lines of evidence in order to best understand the patterns in plant remains that they are encountering within the broad societies that they study. The focus on lived lives is presented in discussions of changing food practices, and considerations of how social relationships such as gender and status altered the roles that agents played in the past. This volume highlights how human interactions with the plant world must be embedded within cultural discussions of food, place, ritual, and gender. Taken together, these paleoethnobotanical case studies move forward theoretical discussions regarding the domestication and spread of plant agriculture, the importance of wild plant taxa in complex societies, landscapes and taskscapes, the meaning and interpretation of ritual, as well as feminist and intersectional approaches to the past.

Additionally, the chapters in this volume reflect the diversity of plant remains that contribute to general paleoethnobotanical practice today. Several chapters draw upon macro-botanical remains including wood, seeds, and storage tissues, while many chapters also incorporate micro-botanical remains including pollen, phytoliths, and starch grains. While all of these datasets contribute to our understanding of species present in past human activities, the contexts in which they are recovered and the variation with which they appear shed light on the roles the plants played in the past. The richest interpretations come from projects that are able to consider the widest range of data types, particularly if we aim to get beyond simple descriptions of food items and environmental settings. Finally, these papers illustrate that plant remains are best interpreted in relation to their associated artifacts and features, making them truly cultural artifacts (Morehart and Morell–Hart 2015).

Paleoethnobotanical analyses provide excellent examples of selection processes, niche construction, resilience, and adaptation, but as anthropologists of the past we also aim to learn about the subtleties of "the hold life has" on these unique places and times from our datasets (Allen 1988: 22). As can be seen with the *Chenopodium* example presented by Fritz and colleagues, we can observe very similar, if not identical, biological processes taking place thousands of miles apart. The experiences and outcomes of those processes, however, differ greatly in their cultural and political context. Rather than attempting to interpret archaeological data through the lenses of one particular theoretical approach, these papers demonstrate how paleoethnobotanical data can, in fact, help rupture traditional patterns of thought and build theory by highlighting the subtle patterns of human behavior that often reveal inconsistencies or surprises that do not fit neatly into specific expectations outlined by particular paradigms. If plant data teach us anything, it is that we should constantly be critical of our assumptions, view the data in new ways, and rather than ignoring irregularities, pursue them and allow them to refine our thinking and model-building.

By bringing to light social interpretations of archaeological plant remains, paleoethnobotanists can make substantial contributions to a wider range of archaeological and anthropological theories and practice. The knowledge that we gain from these investigations is well beyond laundry lists of taxa gathered, and in the moment of interpretation, disparate parts of the past come together to form a broad and at times cohesive vision of lives lived. This tacking back and forth between data and theory is vital to all aspects of archaeology but especially with regard to the social implications of landscape management, the gastropolitical aspects of food, the culturally specific meanings of plants, and the roles that women and men played in constructing these worlds. Paleoethnobotany, therefore, should and will play a key role in the future development of archaeological social theory.

References

Allen, C. (1988). *The hold life has: Coca and cultural identity in an Andean community*. Washington D.C.: Smithsonian Institution Press.

Anderson, M. K. (2000). California Indian horticulture: Management and use of redbud by the Southern Sierra Miwok. In P. E. Minnis (Ed.), *Ethnobotany: A reader* (pp. 29–40). Norman: University of Oklahoma Press.

Arnold, J. E. (Ed.). (1996). *Emergent complexity: The evolution of intermediate societies* (Archaeological Series, Vol. 9). Ann Arbor: International Monographs in Prehistory.

Asouti, E., & Austin, P. (2005). Reconstructing woodland vegetation and its exploitation by past societies, based on the analysis and interpretation of archaeological wood charcoal macro-fossils. *Environmental Archaeology, 10,* 1–18.

Asouti, E., & Fuller, D. Q. (2013). A contextual approach to the emergence of agriculture in southwest Asia. *Current Anthropology, 54*(3), 299–345.

Atalay, S., & Hastorf, C. A. (2006). Food, meals, and daily activities: Food habitus at Neolithic Çatalhöyök. *American Antiquity, 71*(2), 283–319.

Balée, W. (1994). *Footprints of the forest: Ka'apor ethnobotany—The historical ecology of plant utilization by an Amazonian people*. New York: Columbia University Press.

Balée, W., & Erickson, D. (2006). Time and complexity in historical ecology. In W. Balée & D. Erickson (Eds.), *Time and complexity in historical ecology* (pp. 1–17). New York: Columbia University Press.

Bruno, M. C. (2009). Practice and history in the transition to food production. *Current Anthropology, 50*(5), 703–706.

Bruno, M. C. (2014). Beyond raised fields: Exploring farming practices and processes of agricultural change in the ancient Lake Titicaca Basin of the Andes. *American Anthropologist, 116*(1), 1–16.

Crumley, C. L. (1994). Historical ecology: A multidimensional ecological orientation. In C. L. Crumley (Ed.), *Historical ecology: Cultural knowledge and changing landscapes* (pp. 1–16). Sante Fe, NM: School for American Research Press.

Delcourt, H. R., & Delcourt, P. A. (1988). Quaternary landscape ecology: Relevant scales in space and time. *Landscape Ecology, 2*(1), 23–44.

Denham, T. (2005). Envisaging early agriculture in the highlands of New Guinea: Landscapes, plants, and practices. *World Archaeology, 37*(2, Garden Agriculture), 290–306.

Denham, T. (2009). A practice-centered method for charting the emergence and transformation of agriculture. *Current Anthropology, 50*(5), 661–667.

Descola, P. (1994). *In the society of nature: A native ecology in Amazonia.* New York: Cambridge University Press.

Fairbairn, A. (2008). Beyond economy: Seed analysis in landscape archaeology. In B. David & J. Thomas (Eds.), *Handbook of landscape archaeology* (pp. 442–450). Walnut Creek, CA: Left Coast Press Inc.

Ford, R. I. (1979). Paleoethnobotany in American archaeology. *Advances in Archaeological Method and Theory, 2,* 285–336.

Ford, R. I. (1994). Ethnobotany. In R. I. Ford (Ed.), *The nature and status of ethnobotany* (2nd ed., pp. viii–xxxii). Ann Arbor: Museum of Anthropology, University of Michigan.

Fowler, C. S. (1996). Historical perspectives on Timbisha Shoshone land management practices, Death Valley, California. In E. J. Reitz, L. A. Newsom, & S. J. Scudder (Eds.), *Case studies in environmental archaeology* (pp. 87–101). New York: Plenum Press.

Fritz, G. J. (1990). Multiple pathways to farming in pre-contact eastern North America. *Journal of World Prehistory, 4*(4), 387–435.

Gremillion, K. J. (2002). Foraging theory and hypothesis testing in archaeology: An exploration of methodological problems and solutions. *Journal of Anthropological Archaeology, 21*(2), 142–164.

Gremillion, K. J. (2015). Human behavioral ecology and paleoethnobotany. In J. M. Marston, J. D'Alpoim Guedes, & C. Warinner (Eds.), *Method and theory in paleoethnobotany* (pp. 339–354). Boulder: University Press of Colorado.

Hastorf, C. A. (1988). the use of paleoethnobotanical data in prehistoric studies of crop production, processing, and consumption. In C. A. Hastorf & V. Popper (Eds.), *Current paleoethnobotany: Analytical methods and cultural interpretations of archaeological plant remains* (pp. 119–166). Chicago: University of Chicago Press.

Hastorf, C. A. (1990). The effect of the Inka state on Sausa agricultural production and crop consumption. *American Antiquity, 55,* 262–290.

Hastorf, C. A. (1991). Gender, space and food in prehistory. In J. Gero & M. Conkey (Eds.), *Engendering archaeology: Women and prehistory* (pp. 132–159). Oxford: Blackwell Publishers.

Hastorf, C. A. (1993). *Agriculture and the onset of political inequality before the Inka.* Cambridge: Cambridge University Press.

Hastorf, C. A. (1998). The cultural life of early domestic plant use. *Antiquity, 72,* 773–782.

Hastorf, C. A. (1999). Recent research in paleoethnobotany. *Journal of Archaeological Research, 7*(1), 55–103.

Hastorf, C. A. (2001). Making the invisible visible: The hidden jewels of archaeology. In P. Drooker (Ed.), *Fleeting identities: Perishable material culture in archaeological research* (pp. 27–42, Center for Archaeological Investigations Occasional Paper 28). Carbondale, IL: Southern Illinois University Press.

Hastorf, C. A. (2003). Andean luxury foods: Special food for the ancestors. *Deities and the Élite. Antiquity, 77*(29), 545–554.

Hastorf, C. A. (2006). Domesticated food and society in early coastal Peru. In W. Balée & C. L. Erickson (Eds.), *Time and complexity in historical ccology: studies in the Neotropical lowlands* (pp. 87–126). New York: Columbia University Press.

Hastorf, C. A., & Johannessen, S. (1991). Understanding changing people/plant relationships in the prehispanic Andes. In R. W. Pruecel (Ed.), *Processual and postprocessual archaeologies: Multiple ways of knowing the past* (pp. 140–158). Carbondale, IL: Center for Archaeological Investigations.

Hastorf, C. A., & Johannessen, S. (1996). Understanding changing people/plant relationships in the prehispanic Andes. In R. Preucel & I. Hodder (Eds.), *Contemporary archaeology in theory: [a reader]* (pp. 61–78). Oxford: Blackwell.

Hastorf, C. A., & Popper, V. (1989). *Current paleoethnobotany: Analytical methods and cultural interpretations of archaeological plant remains.* Chicago, IL: University of Chicago Press.

Ingold, T. (1996). The optimal forager and economic man. In P. Descola & G. Pálsson (Eds.), *Nature and society: Anthropological perspectives* (pp. 25–44). New York: Routledge.

Ingold, T. (2000). *The perception of the environment: Essays on livelihood, dwelling and skill*. New York: Routledge.

Johannessen, S., & Hastorf, C. A. (1990). A history of fuel management (A.D. 500 to the present) in the Mantaro Valley, Peru. *Journal of Ethnobiology, 10*(1), 61–90.

Jones, M. K. (1985). Archaeobotany beyond subsistence reconstruction. In M. K. Jones (Ed.), *Beyond domestication in prehistoric Europe* (pp. 107–128). Oxford: Oxbow Press.

Jones, V. (1941). The nature and status of ethnobotany. *Chronica Botanica, 6*(10), 219–221.

Lentz, D. L. (1991). Maya diets of the rich and poor: Paleoethnobotanical evidence from Copan. *Latin American Antiquity, 2*(3), 269–287.

Marston, J. M., Warinner, C., & D'Aploim Guedes, J. (2015). Paleoethnobotanical method and theory in the twenty-first century. In J. M. Marston, J. D'Aploim Guedes, & C. Warinner (Eds.), *Method and theory in paleoethnobotany* (pp. 1–16). Boulder: University Press of Colorado.

Morehart, C. T., & Helmke, C. G. B. (2008). Situating power and locating knowledge: A paleoethnobotanical perspective on Late Classic Maya gender and social relations. In E. Brumfiel, & C. Robin (Eds.), *Gender, households and societies: Unraveling the threads of the past and the present*. Archaeological Papers of the American Anthropological Association *18* (1), 60–75. New York: Wiley-Blackwell.

Morehart, C. T., Lentz, D. L., & Prufer, K. M. (2005). Wood of the gods: The ritual use of pine (*Pinus* spp.) by the ancient lowland Maya. *Latin American Antiquity, 16*(3), 255–274.

Morehart, C. T., & Morell-Hart, S. (2013). Beyond the ecofact: Toward a social paleoethnobotany in Mesoamerica. *Journal of Anthropological Archaeology, 22*(2), 483–511.

Morehart, C. T., & Morell-Hart, S. (2015). Beyond the ecofact: Toward a social paleoethnobotany in Mesoamerica. *Journal of Archaeological Method and Theory, 22*, 483–511.

Morell-Hart, S. (2015). Paleoethnobotanical analysis, post-processing. In J. M. Marston, J. D'Aploim Guedes, & C. Warinner (Eds.), *Method and theory in paleoethnobotany* (pp. 371–390). Boulder: University Press of Colorado.

Pearsall, D. M. (2015). *Paleoethnobotany: A handbook of procedures* (3rd ed.). Walnut Creek, CA: Left Coast Press.

Piperno, D. R., & Pearsall, D. M. (1998). *The origins of agriculture in the lowland Neotropics*. San Diego: Academic Press.

Reitz, E. J., Newsom, L. A., Scudder, S. J., & Scarry, C. M. (2008). Introduction to environmental archaeology. In E. J. Reitz, C. M. Scarry, & S. J. Scudder (Eds.), *Case studies in environmental archaeology* (2nd ed., pp. 3–19). New York: Springer.

Renfrew, J. (1973). *Paleoethnobotany: The prehistoric food plants of the Near East and Europe*. New York: Columbia University Press.

Rival, L. (Ed.). (1998). *The social life of trees: Anthropological perspectives on tree symbolism*. Oxford: Berg.

Rosen, A. M. (2007). *Civilizing climate: Social responses to climate change in the ancient Near East*. Lanham, MD: Alta Mira Press.

Sayre, M. (2014). Ceremonial plants in the Andean region. In A. Chevalier, E. Marinova, & L. Peña-Chocarro (Eds.), *Crops and people: Choices and diversity through time* (pp. 369–373). Brussells: European Science Foundation/Early Agricultural Remnants and Technical Heritage Program.

Smart, T. L., & Hoffman, E. S. (1988). Environmental interpretation of archaeological charcoal. In C. A. Hastorf & V. Popper (Eds.), *Current paleoethnobotany: Analytical methods and cultural interpretations of archaeological plant remains* (pp. 167–205). Chicago: University of Chicago Press.

Smith, B. D. (2007). Niche construction and the behavioral context of plant and animal domestication. *Evolutionary Anthropology, 16*(5), 188–199.

Smith, B. D. (2011). A cultural niche construction theory of initial domestication. *Biological Theory, 6*(3), 260–271.

Smith, B. D. (2015). Documenting human niche construction in the archaeological record. In J. M. Marston, J. D'Alpoim Guedes, & C. Warinner (Eds.), *Method and theory in paleoethnobotany* (pp. 355–370). Boulder: University Press of Colorado.

VanDerwarker, A., Bardolph, D., Hoppa, K., Thakar, H. B., Martin, L. S., Jaqua, A. L., et al. (2016). New World paleoethnobotany in the new millenium (2000–2013). *Journal of Archaeological Research, 24*(2), 125–177.

Watson, P., & Kennedy, M. (1991). The development of horticulture in the Eastern Woodlands: Women's role. In J. Gero & M. Conkey (Eds.), *Engendering archaeology: Women and prehistory* (pp. 255–275). Oxford: Blackwell.

Chapter 2
Integrating Archaeological Data Toward a Better Understanding of Food Plants Choices and Territory Exploitation in the Northwestern European Early Neolithic: The Case of Remicourt "En Bia Flo II"

Alexandre Chevalier and Dominique Bosquet

The Belgian Early Neolithic (LBK) Within the Overall European Neolithization Process

Initiated at the Pleistocene-Holocene transition, around 13,000 years ago in the Near East, the neolithization process took some 6000 years to reach Northwestern Europe, around 5300 BCE through the Danubian expansion (Mazurié de Keroualin 2001). The *Linearbandkeramik* (LBK) is the Danubian central and northwestern cultural component that presents a ceramic, lithic and architectural homogeneity that spreads across more than 2000 km from Northern Serbia to Middle Belgium and the Paris basin, across Hungary, Austria, Slovakia, the Czech Republic, Poland, Germany, Eastern France, and The Netherlands (Fig. 2.1).[1] The Belgian LBK is therefore situated on the western edge of an expansion area which spread from the Black Sea to Normandy in its maximum extension (Golitko 2015: 33, Fig. 2.4). From a chronological point of view, it seems that the LBK expands westward within a relatively short period of time, roughly from 5600 BCE to 4700 BCE (Golitko 2015; Jadin 2003). The LBK was not an in situ neolithization process, but

[1]The two other main contemporary European Neolithic expansion traditions are the Cardial in the Adriatic Sea area, Italy, Southern France and eastern Spain; and the Chasséen in Central and Western France, and the British Islands.

A. Chevalier (✉)
Royal Belgian Institute of Natural Sciences, Brussels, Belgium
e-mail: achevalier@naturalsciences.be

D. Bosquet
SPW-DGO4-Service de l'Archéologie en province de Brabant Wallon, Wavre, Belgium
e-mail: dominique.bosquet@spw.wallonie.be

M.P. Sayre and M.C. Bruno (eds.), *Social Perspectives on Ancient Lives from Paleoethnobotanical Data*, DOI 10.1007/978-3-319-52849-6_2

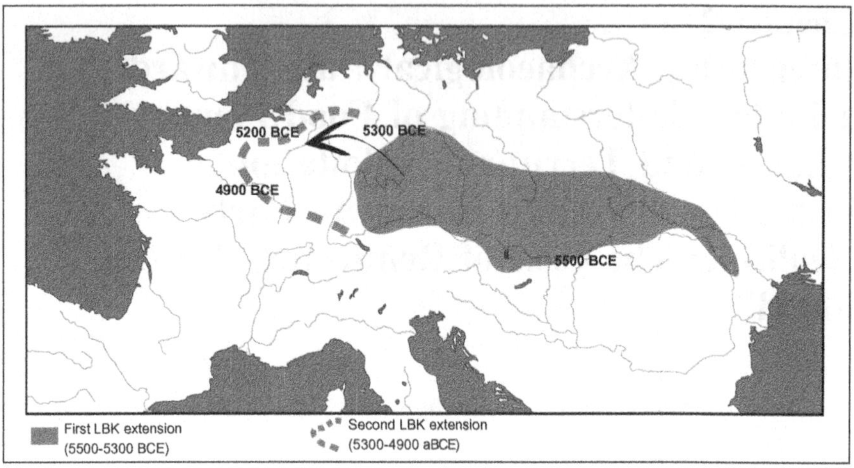

Fig. 2.1 Map of the LBK extension (adapted from Salavert 2010a)

a population movement—it is still a matter of debate whether it was a conscious migration or demographic expansion, or both—that brought along domesticated plant and animals, introduced agricultural and pastoral practices, as well as ceramic technology. It initially began in forested areas growing on loess soils, then spread into other ecological settings, most of the time simply replacing nomadic Mesolithic groups. With only a few exceptions, such as the Limbourg group in current Central Benelux, or the Blicky group in Southern Belgium, no interaction or acculturation processes were observed between hunter-gatherers and agro-pastoral groups (Mazurié de Keroualin 2001; Gallay 1995). Mesolithic groups were most likely displaced in less productive areas, where their lifestyle faded under the agro-pastoral activities rather than integrated into the new Neolithic populations. Agro-pastoral groups eventually replaced completely nomadic and semi-nomadic Mesolithic hunter-gatherers within about five centuries, for there is no more evidence of the Mesolithic tradition by the beginning of the fifth millennium in Northwestern Europe (ibid.).

Economic practices imply specific mindsets about social relationships, space organization, and perception of nature (Ingold 2000): the switch from a predation economy to an exploitation economy implies, therefore, strong changes in social and spatial organizations, as well as the way humans identify themselves with nature.

Two Theoretical Approaches to Interpreting the Transition from Predation to Food Production

Of the many theoretical schools trying to understand and predict human–nature relationships in archaeology, Human Behavioral Ecology (HBE), and Historical Ecology (HE) are the two predominant ones currently used to explain the neolithization process (domestication and/or spread of domesticated plants). Niche Construction Theory (NCT), which bridges some of the differences between HBE and HE, and has recently been applied to explain domestication processes around the world and human resilience to ecological change (Smith 2012, 2015; Boivin et al. 2016). We briefly present here HBE and HE only, first because we are not faced with a domestication process *sui generis*, and second because while NCT, unlike HBE, recognizes the importance of mutual interactions of the co-evolutionary process between human groups and nature, it fails to recognize that cultural inheritance is equally as important as ecological and genetic ones. Cultural processes are often the trigger of changes, or the origin of a status quo, which induce ecological and genetic inheritances, as HE explicitly recognizes.

HBE, also called evolutionary ecology, aims at predicting behavioral responses to changing socio-ecological conditions, such as: exceeding environmental carrying capacity due to demographic increase; altering living resources, which either leads to drastically limiting them or to redistributing them; climatic changes; and social change via endogenous socioeconomic competition, which can happen, according to HBE, only in resource-rich areas and in societies with prior socioeconomic differentiation. In order to understand these responses and to predict them, HBE uses one main assumption: humans tend to optimize their actions (inputs) to get the most out of them (outputs). The different models that HBE uses are based on economic concepts, such as marginal value, opportunity costs, discounting, or risk analysis. The use of these concepts implies that humans behave rationally and that they assess the risks and benefits "of alternative courses of action under a range of environmental conditions" (Winterhalder and Kennett 2006: 13) when they decide which course of action they will take.

Several models predicting specific human behaviors stem from these main assumptions and economic concepts: *diet breadth* implies a ranking of resources and their frequency in a given space in order to make predictions about whether past groups exploited specific resources or not; *patch choice predictability* depends on the concentration of a given resource and the ranking of the resource: *patch residence time* predicts that the bigger, and therefore the longer, the travel time across it is, the longer the residence time will be; *habitat selection*, an habitat may enclose several patches, prediction is driven by both the abundance of given resources and the density of population based on the concept of Ideal Free Distribution, that is an optimal occupation of habitats according to their ranking and the chronology of settlements/settlers on a first come-first serve basis; *central place foraging* predicts that the more costs, both in travel time and in distance, are associated to reach a resource, the more this resource must be valuable either in quantities or in quality; *settlement relocation* aims at predicting when groups must change their *central place* because of depleted resources, their seasonal availability or a change in their assigned value, taken into account the cost associated to move from one location to another one (opportunity costs).

All these different models have been widely used in archaeology (Kennet and Winterhalder 2006; Smith and Winterhalder 1992), and in particular in the Aisne valley in Northern France, to explain the distribution of LBK sites and the exploitation of the territory by their settlers (Dubouloz et al. 2012). It is not the place to criticize here the HBE theories; there is enough literature showing potential flaws in reasoning and application (Ingold 1996; Zeder 2012). However, we would like to stress that research in microeconomics and consumer behavior has been trying for decades to understand and to predict current consumer's choices based on the same concepts used by HBE, without any clear and definitive success; consumer's choices tend not to follow the microeconomic models (Mason 1998). In fact, consumer choices may appear illogical and irrational if we do not take into account their individual social and psychological settings (Achar et al. 2016; Foxall 2009), which economic models rarely did until recently. So how can we apply the very same concepts and pretend to predict behaviors of past human groups? In particular, HBE neglects to include social and cultural factors that influence food choices over time, for in human groups these factors are crucial when it comes to selecting food from a vast array of possibilities, and to decide which ones are good to eat and which ones should be avoided (Chevalier et al. 2014a).

Historical ecology aims at showing how "human societies [...] rather than adapt their subsistence activities, seasonal rounds populations sizes, settlement sizes, and so on to preexisting constraints of the environment—have transformed most of those constraints into negligible phenomena as concern species diversities, landscape heterogeneity, archaeological site formation, and the development of the built environment more generally" (Balée and Erickson 2006: x). As such, Historical Ecology opposes the concepts and models of HBE (and goes beyond NCT); humans are not objects of the environments who need to continuously adapt to its changes. On the contrary, they are considered as agents of these changes, not only in the sense of degradation, but most of it in the sense of enhancement of its total biological richness and landscape diversity. "Environments are […] adapted to the sociocultural and political systems (or to humans' needs and desires) that have coexisted with them" (ibid: 4). In turn these modifications influence human groups, their perceptions and their sociocultural organization; and so on in an endless movement of reciprocal influence.

The central object of Historical Ecology is the landscape as both witness of the many activities humans carried on in the biosphere and as a building agent of human sociocultural and political systems: it is the book of human history which is "embedded in the local and regional landscape" (ibid.: 6), that can be read through "pattern of residues, anomalies and cultural imprints" (ibid.: 7) translated into biodiversity, biogeography, archaeobotany, geology, soil chemistry, linguistic, or material culture studies.

Conscious transformation of the landscape, by favoring specific plant species for human purposes, has brought modifications that are observable in forest tree composition, as described by Balée (1994) for the Amazon basin, for instance, and between the pioneer LBK phase and the second occupation phase of occupation in Hesbaye, to give room for agriculture and pasture, as we will show with pollen and wood charcoal analyses. The analysis of botanical macroremains, of course, reveals information regarding the introduction of domesticates, but weeds, as we will show, are of particular interest, not only to study the switch from a predation to a production economy, but also to assess agricultural practices.

Analytical Tools of Environmental Archaeology and Foodways

Several analytical tools can provide information for paleoecological and past land cover reconstitutions, such as pollen and wood charcoal analyses. However, obtaining representative data sets requires new approaches for obtaining such information, for pollen grains are not preserved in most types of soil. In addition, this proxy can only provide regional and supraregional information. When it comes to understanding the local human impact on vegetation, another proxy is needed together with wood charcoal analysis, namely phytolith analysis, in order to obtain accurate data on vegetation cover changes on the small patches surrounding settlements.

Similarly, the dominant method in archaeology to analyze plant choice for food is macrobotanical analysis (Hastorf 1990, 1993, 2003). Unfortunately, analysis of seeds remains is extremely limited in this region mainly because the soils of this region are not amendable to their preservation. Moreover, plants that produce items such as tubers simply cannot be identified in archaeological sites because they do not leave remains once ingested as food or medicine. Microbotanical remains such as phytoliths and starch grains are therefore necessary to analyze when soil conditions or age deposits do not permit other plant remains to be preserved. These microremains are complementary to other archaeobotanical analyses in normal conditions of preservation that permit archaeologists to get information on plant parts that may not leave macroremains.

Different lines of evidence—seed, wood charcoal, phytoliths, and starch grain analyses—have been applied to several sites in the Hesbaye region in order to better understand the human–nature relationships and social processes that took place during initial LBK settlement and later periods of the early Neolithic occupation (Table 2.1). In this article, we focus on the site of Remicourt "En Bia Flo II" with some comparisons to other sites when data are available. Phytoliths, starch grains, and pollen analyses are ongoing and will be applied in the future to other LBK settlements in order to broaden and refine the information already available on changes in vegetation cover during the Belgian LBK (Heim 1985), as well as in food procurement processes and social organization.

Table 2.1 Hesbaye sites with the analyses carried on and their respective C14 dates

Sites	Pioneer houses	Later phases houses	Seeds analysis	Wood charcoal analysis	Pollen analysis	Starch grain analysis	Phytolith analysis	C14 dates[a]
Remicourt "En Bia Flo II"	1	9	Yes	Yes	No	Yes	Yes	6108–6219 BP
Remicourt "Fond de Momalle"	1 or 2	11 or 10	No	No	No	No	Yes	–
Remicourt "Tombe de Hodeige"	–	–	No	No	No	No	Yes	–
Fexhe-le-Haut-Clocher "Podrî l'Cortri"	3	9	Yes	Yes	No	No	Yes	6120–6240 BP
Fexhe-le-Haut-Clocher/Voroux-Goreux	–	–	No	No	No	No	Yes	–
Waremme "Longchamps"	1	1	Yes	Yes	Yes	No	No	–
Darion "Colia"	1	4	Yes	Yes	Yes	No	No	6060–6205 BP
Alleur "Domaine militaire"	–	3	Yes	Yes	No	No	No	–

[a]Only AMS dates on short life samples (*Triticum* spp., charred hazelnuts); dates on charcoal have been excluded

Belgian LBK

The Belgian LBK was identified during the late 19th century, but it was only thoroughly documented beginning in the 1960s with the first extensive excavations at the site of Rosmeer (Roosens 1961). These excavations were followed in the years 1980–1990 by a series of excavations in three contiguous Belgian provinces: in the Hesbaye in the Liège province, where more than 200 sites are known (Cahen et al. 1990), in Flemish Brabant (Lodewijckx 1990) and in the western Hainaut (Constantin and Demarez 1984). Thanks to archaeological rescue operations undertaken between 1993 and 1999 by the Walloon region, together with the Royal Belgian Institute of Natural Sciences, an unprecedented corpus of sites has been uncovered and made available for study (Bosquet et al. 2004; Jadin 2003; Keeley et al. 2005; Livingstone-Smith et al. 2012) on the high-speed train path between the French border and the town of Liège (Fig. 2.2). The University of Illinois (Chicago,

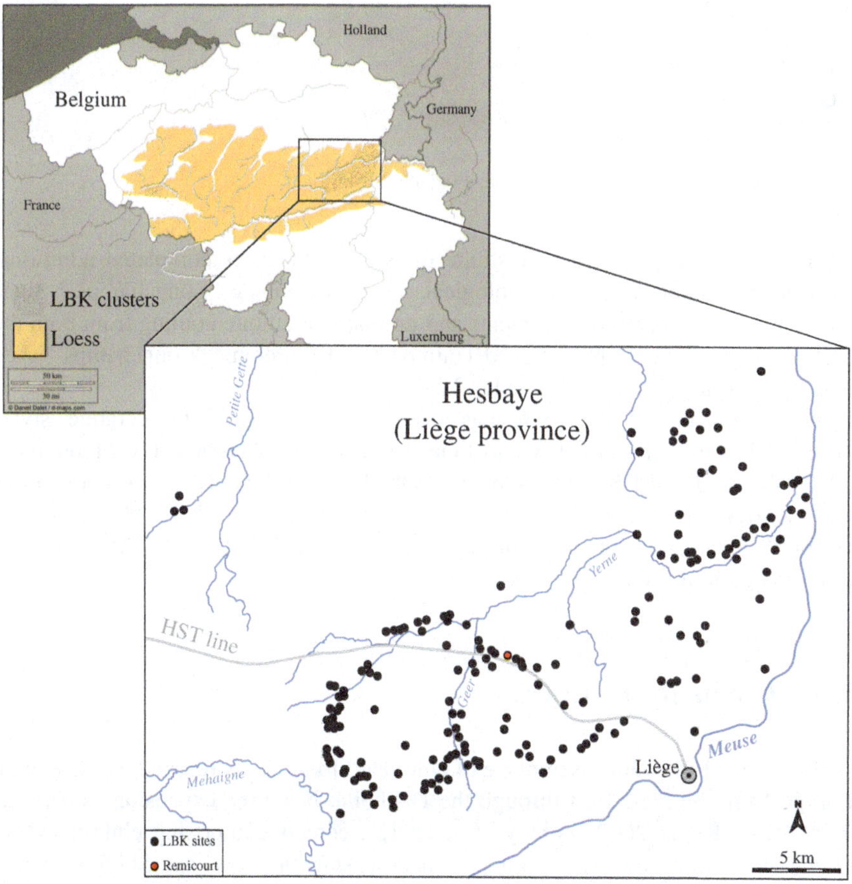

Fig. 2.2 Map of Belgium with the LBK sites cluster

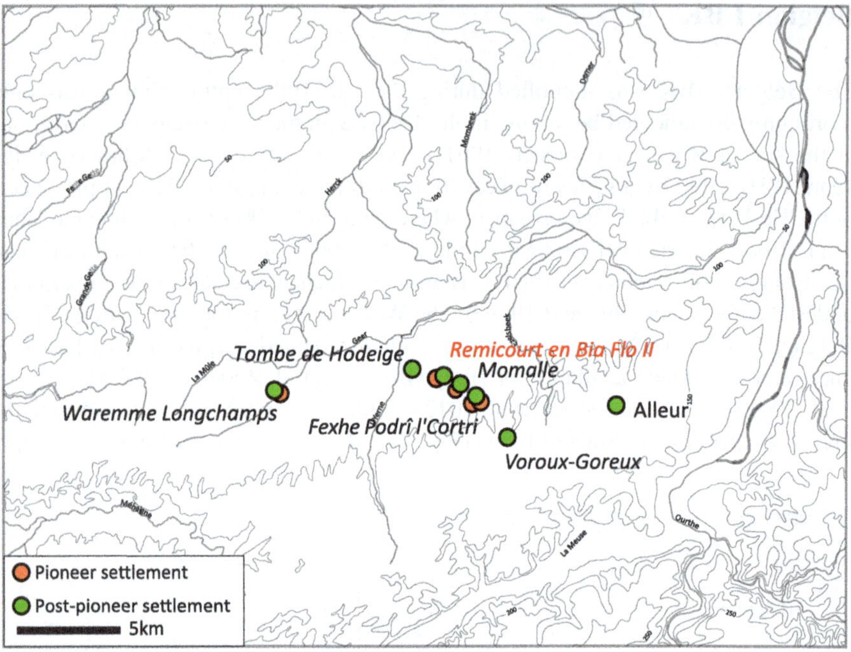

Fig. 2.3 Map of the LBK sites in Hesbaye (adapted from Salavert 2010a)

USA) and the Royal Belgian Institute of Natural Sciences completed additional work in the region at the archaeological site of Waremme "Longchamps." Since 2000, ongoing research on sediments and on material culture coming from both the LBK sites found on the High-Speed Train path and Waremme "Longchamps", shed new light on the Belgian LBK (Fig. 2.3).

A subdivision of the LBK has been proposed based on ceramic styles (Table 2.2)—from Earliest LBK to Final LBK, (Meier-Arendt 1966; Modderman 1970): the Belgian LBK belongs mainly to the Late and Final LBK, with only a few dates associated with the Early LBK (Bosquet and Golitko 2012; Bosquet et al. 2008). A local, more precise, chronology (following Blouet et al. 2013 methodology) is currently under development.

LBK Foodplant Practices

A very limited standard assemblage of cultivated food plants reached Northwestern Europe from the Near East through the Danubian Neolithic expansion as early as 5500 BCE (Kreuz 2007; Zohary et al. 2012): cereals comprising einkorn wheat (*Triticum monococcum* L.), emmer wheat (*Triticum dicoccum* Shübl.), barley (*Hordeum vulgare* L., including naked and hulled and 2- and 6-row varieties);

Table 2.2 Comparative chronologies of the LBK in Europe based on ceramic styles (adapted from Salavert 2010a)

	Germany (Hesse) (Meier-Arendt 1966)	Rhine-Meuse Group (Modderman 1970)	Paris Basin (Constantin & Ilett 1997)	
				4900 BCE
Final LBK		II d+	RRBP	
		II d		
Late LBK		II c		
	LBK II-V	II b		
		II a		
Middle LBK		I d		
		I c		
Early LBK		I b		
				5300 BCE
Earliest LBK	LBK I			

legumes such as peas (*Pisum sativum* L.) and lentils (*Lens culinaris* Medik.), but in quite low quantities; and finally flax (*Linum usitatissimum* L.). In addition to these cultivars, many edible wild plants are found in every archaeological site dating to this period, the precise representation of which depends on the specific ecologies surrounding these sites—the most common are hazelnut (*Corylus avellana* L.) and apple (*Malus* spp.). Northwestern European archaeological sites also contain other plants including rare remains of poppy (*Papaver somniferum* L.), free-threshing wheat (*Triticum aestivum* L./*durum* Desf./*turgidum* L.), rye (*Secale cereale* L.), millet (*Panicum miliaceum* L.), bitter vetch (*Vicia ervilia* L.), and Celtic bean (*Vicia faba* L.). Because their occurrence is very low, these food plants may not constitute real cultivars, but rather weeds incidentally grown with other crops. However, this is a matter of debate, as more evidence for these species are found through microfossil analyses (Chevalier and Bosquet 2013).

The distribution of these crops does not show any clear ecological pattern and cannot be related to only different soil or climatic conditions (Kreuz 2007), which would allow us to use HBE to interpret these data. Cultural differences seem to be the primary explanation for the variation in crop distribution across Central and Northwestern Europe, and therefore, HE provides a more appropriate framework with which to understand these processes.

Remicourt "En Bia Flo II" Botanical Data

This archaeological site is located in the village of Momalle, in the municipality of Remicourt, in a place named "*En Bia Flo*" at 130masl. It was excavated between 1997 and 1998 over an area of 6847 m^2 (Bosquet and Preud'Homme 1998). The site was comprised of 235 structures distributed across two sectors: a partially excavated village of at least 10 houses surrounded by an enclosure composed of a deep ditch and a wooden palisade with an opening on its eastern side divided by a *schlitzgrube*,[2] and a house located 130 m away from the village, facing the fence opening (Fig. 2.4). Among the structures there are 182 pits whose primary function is unknown, but most of them were ultimately used as waste dumps and contained charred botanical remains, flint debitage, sickle blades, ceramic sherds, and querns.

The village was initially dated to the Late and Final LBK according to ceramic typology, while the outlying house was dated to the Middle and Late LBK. This chronology has been confirmed by C14 dates (Fig. 2.5): the isolated house dates from 5301 to 4990 BCE at 2σ, and the village dates from 5286 to 4941 BCE at 2σ. This specific chronological and geographical configuration allows us to propose the existence of a pioneer settlement phase of the region and/or the presence of buildings whose particular social status explains their isolation from their associated villages (Bosquet et al. 2004; Bosquet and Golitko 2012; Salavert 2010a).

We do not yet have direct pollen analyses for the site of Remicourt "En Bia Flo II", but, according to (Heim 1985), the deepest soil layers in different pits at the nearby LBK site of Darion show a pollen composition representing a dense forest (between ~87 and 74% of arboreal pollen, or "AP") of *Ulmus* spp. (elm), together with *Corylus avellana* subsp. (hazel) and *Alnus* spp. (alder) species. Non-AP pollen were represented mostly by Poaceae (grass), as well as by ferns and mosses, which was expected in a forest cover setting. In the middle layers of the pits, arboreal pollen taxa were still dominant, but the forest was already more open, and was dominated by *Tilia* spp. (linden) and *Corylus avellana* subsp. (hazel), with a strong presence of *Hedera helix* L. (ivy), a plant that needs full exposure of the sun to grow. Non-arboreal taxa were the same as in the deepest layer, with a dominance of

[2]A Schlitzgrube is a ditch with a profile in "V", that appears with the LBK and is usually associated with the Neolithic. This kind of structure may have a specific signification in particular when associated with a village gate and therefore potentially specific remains.

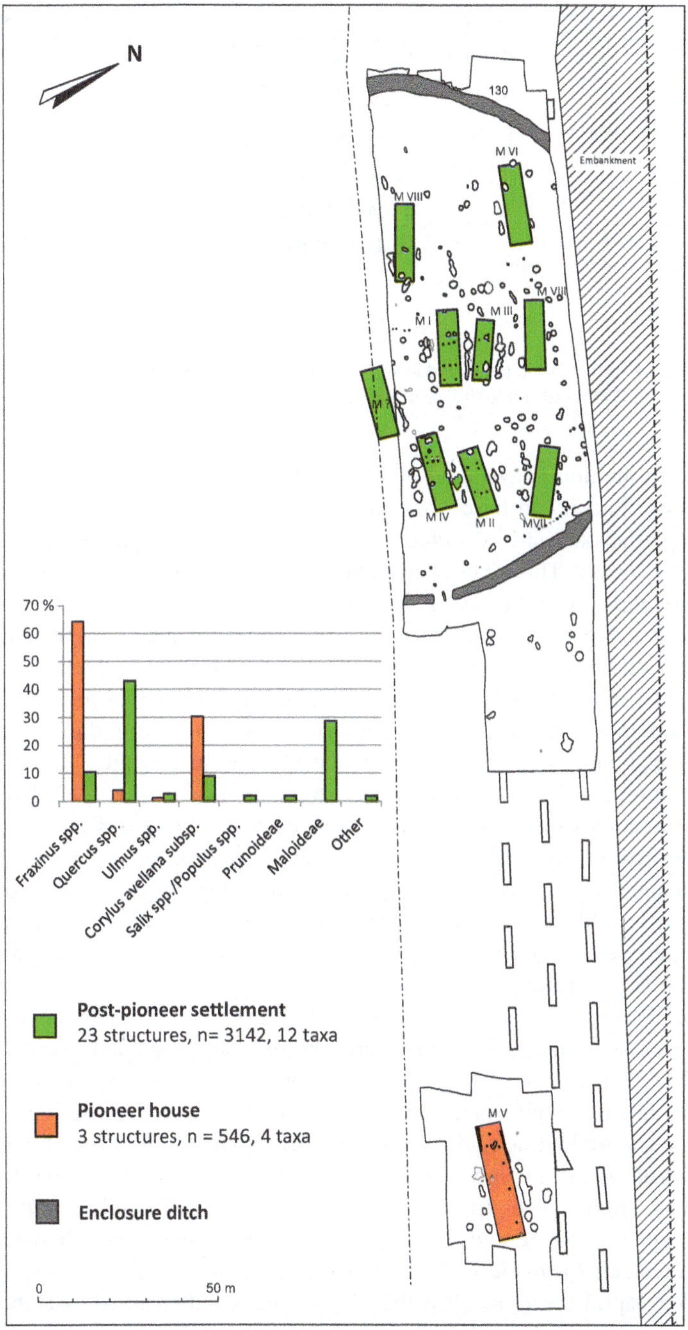

Fig. 2.4 Map of the excavated parts of the site of Remicourt "En Bia Flo" and wood charcoal species repartition (pioneer house vs. post-pioneer settlement)

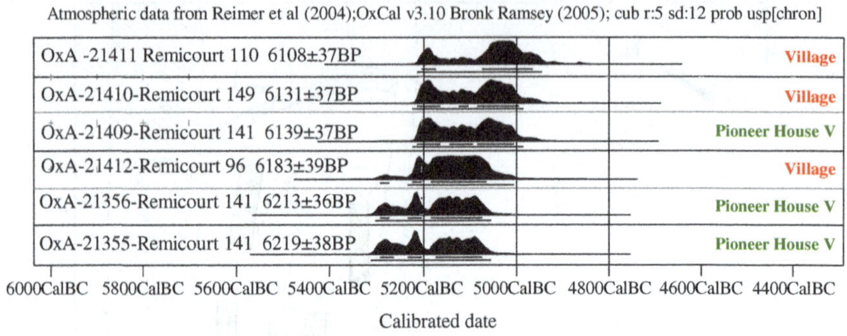

Fig. 2.5 Graph of the calibrated C14 dates for the site of Remicourt "En Bia Flo II" (OxCal v3.10 Bronk Ramsey, 2005; cub r:5 sd:12 prob usp[chron])

Cerealia. In the upper layers, arboreal taxa dropped down to 44%, with the same composition: linden and hazel. Non-arboreal taxa were mostly represented by the *Crepis* type (hawksbeard), *Calluna vulgaris* L. (Hull.) (heather), Poaceae (grasses) as well as *Cerealia*. This plant composition reflects an open space, with soil erosion and acidification, since *Calluna* is linked to acid soils, which is not the case for the loessic soils in the Hesbaye. *Cerealia* pollen is less abundant in some contexts than in the previous phase, but this is a hedge effect of *Corylus*, which colonized the former forest spaces. Finally, the dominance of *Tilia* spp. could be either linked to specific activities or to the climatic conditions prevailing during the Atlantic phase.

Wood charcoal analyses have been carried out for 26 pits,[3] three from the pioneer house, and 23 from post-pioneer structures (Salavert et al. 2014). Most of the material comes from layers described as classical refuse pits and consists of a mix of everyday life waste, such as ceramic sherds, lithic industry, charred macroremains (wood charcoal, seeds, fruits), and burned clay or daub fragments. These detritic sublayers may represent several events of fuel collection, and charcoal assemblages may correspond to the wood vegetation in the fuel supplying area around the site (Chabal 1994).

In the three pioneer pits, 546 charcoal fragments were identified (Fig. 2.4, Table 2.3). The main taxa identified were *Fraxinus* spp. (ash) and *Corylus avellana* subsp. (hazel). Fragments of *Quercus* spp. (oak) and *Ulmus* spp. (elm) were also identified. In the 23 post-pioneer pits, 3142 fragments were identified. *Quercus* spp. (oak), and Maloideae (24%) (apple, medlar, or hawthorn subfamily) dominated together with *Fraxinus* spp. (ash) and *Corylus* avellana subsp. (hazel). *Ulmus* spp. (elm), Prunoideae (prune, cherry or peach subfamily), *Salix/Populus* (willow/poplar), and *Sambucus* spp. (elder) were also present (between 1 and 3%), whereas cf. *Frangula* (alder), *Tilia* spp. (linden), cf. *Ligustrum* (privet), and *Acer* spp. (maple) comprised less than 1% of the wood charcoal assemblage.

[3]Field and laboratory processes are detailed in Annex 1.

Table 2.3 Wood charcoal identifications for the site of Remicourt "En Bia Flo II" (adapted from Salavert 2014)

	Remicourt "*En Bia Flo II*" (MV)					
	Pioneer occupation			Post-pioneer occupation		
Number of fragments	546			3142		
Number of structures	3			23		
Taxa	n	%	p/a	n	%	p/a
Quercus sp.	22	4	2	1349	42.9	22
Fraxinus excelsior	351	64.3	3	330	10.5	14
Corylus avellana	166	30.4	3	279	8.9	14
Ulmus sp.	7	1.28	2	84	2.7	8
Tilia sp.	–			6	0.2	4
Maloideae	–			901	28.7	22
Prunoideae	–			67	2.1	13
Sambucus sp.	–			43	1.4	7
Cf. *Frangula* sp.	–			12	0.4	3
Cf. *Ligustrum* sp.	–			1	0.03	1
Acer sp.	–			1	0.03	1
Salix/Populus	–			69	6	2.2

Seed identification was carried out on 37 structures (Table 2.4),[4] and 70.8% of all non-woody macroremains were from wild and domesticated cereals (Salavert 2010a, 2011). With the exception of non-identifiable *Cerealia* remains, wheat (*Triticum* sp.) constituted virtually the only cereal processed, with slightly more emmer (*Triticum dicoccum* L.) than einkorn (*Triticum monococcum* L.). In addition, some possible barley remains (cf. *Hordeum vulgare* L.) have been identified. Together with these cereals, hazelnut (*Corylus avellana* L.) as well as wild plants have also been identified; some of these may have been gathered for food, such as goosefoot (*Chenopodium album* L.), nipplewort (*Lapsana communis* L.), black-bindweed (*Fallopia convolvulus*), or rye brome (*Bromus secalinus* L.); some other can be considered as weeds associated either with cereal agriculture such as the brome (*Bromus* spp.), or with animal husbandry and pastures with soils heavily trampled by animals, such as rye/soft brome (*Bromus secalinus/mollis*), willow weed (*Persicaria lapathifolia* L.), and curled dock (*Rumex* cf. *crispus*). Finally, two opium poppy seeds (*Papaver somniferum* L.) were identified (Salavert 2010b).

Phytolith analyses have been carried on only one soil context at Remicourt so far,[5] a sample coming from the *Schlitzgrube* associated with the village's surrounding walls (Fig. 2.6). This sample had a medium concentration of phytoliths and quite a high diversity of forms. It mostly contained morphotypes found in trees

[4]Field and laboratory processes are detailed in Annex 1.

[5]See Annex 1.

Table 2.4 Seed identifications for the site of Remicourt "En Bia Flo II" (adapted from Salavert 2010a)

Remicourt "*En Bia Flo II*"		
Number of structures	37	
Number of deposits	79	
Number of samples	99	
Volume/weight of screened soils (l/kg)	154/64	
Cultivated plants	n	%
Triticum dicoccum—caryopsis	53	6.1
Bases spikelet	22	2.5
Triticum cf. dicoccum—caryopsis	9	1.0
Triticum monococcum—caryopsis	8	0.9
Bases spikelet	4	0.5
Triticum cf. monococcum—caryopsis	7	0.8
Triticum sp.—caryopsis	82	9.4
Bases spikelet	2	0.2
Glumes	36	4.1
Cf. *Triticum* sp.—caryopsis	4	0.5
Cf. *Hordeum* sp.—caryopsis	4	0.5
Cerealia	384	44.2
Papaver somniferum	2	0.2
Weeds and wild plants		
Lapsana communis	11	1.3
Chenopodium album	63	7.2
Fallopia convolvulus	25	2.9
Bromus secalinus/mollis	38	4.4
Bromus tectorum/sterilis	17	2.0
Bromus sp.	22	2.5
Cf. *Bromus* sp.	10	1.2
Rumex cf. *crispus*	8	0.9
Persicaria lapathifolium/mitis	4	0.5
Small Fabaceae	9	1.0
Fruit trees		
Corylus avellana	9	1.0
Herbaceous root	12	1.4
Stem Poaceae	8	0.9
Flower bud	16	1.8
Total number of identified remains	869	

and in Panicoidae plants, which is quite a different composition in comparison with other sites in the Hesbaye region. It is quite surprising since Panicoideae is not a subfamily well represented in Northwestern Europe, with only eight potential

Fig. 2.6 Map of the excavated parts of the site of Remicourt "En Bia Flo" with the location of the sampled pits for phytoliths and starch grains analyses (adapted from Salavert 2010a)

species present during the LBK,[6] because the moist and cool conditions prevailing in Belgium are more favorable to the Pooideae grasses subfamily (Chevalier and Bosquet 2010).

In addition, starch grain analyses,[7] together with phytolith analyses, have been applied to grinding stones in order to identify both the functionality of these querns and the plants ground, and most likely, eaten by the LBK settlers. Nine grinding stones from different pits (Fig. 2.6) were selected among a collection of 32 in order to have a representative distribution of shapes, sizes, and rock composition. There were starch grains on only six grinding stones (#274, 275, 277, 278, 280, 281) and diagnostic phytoliths were found on only five specimens (#274, 275, 277, 278 and 280). Phytolith analyses (Table 2.5) were not very informative; with the exception of grinding stone #277 where we uncovered a small number of very specific forms that could not be associated with any specific taxon, all the querns seemed to have been used to grind various kind of plants, including dicotyledons, but none provided any evidence of cereal processing. In contrast, starch grain analyses yielded very promising results (Table 2.6). To date we have already identified wheat (*Triticum* sp.), barley (*Hordeum* sp.), oat (*Avena* sp.), pea (*Pisum* sp.), and acorn (*Quercus* sp.) granules embedded in the porosities of some of the grinding stones (Fig. 2.7a–h).

Early Neolithic Paleoenvironment in Belgium

The pollen data (Bakels 1992a) clearly indicate a progressive forest clearing, linked with the implementation of cereal agriculture, the necessity to create pastures for large ruminants such as cows, and wood exploitation for fuel, crafts, and buildings. The first LBK settlers consciously transformed the environment to adapt it to their needs, which is in line with HE models; however, these data are imprecise about which plants were cultivated, or which specific cereals were grown and eaten.

Wood charcoal analyses (Salavert et al. 2014) confirm these trends, although with a higher level of resolution. During the pioneer settlement period, only a few forest species were exploited, predominately oak, ash, and hazel trees. In this context, elm could have been associated with the deciduous oak forest or referred specifically to the valley bottoms. The absence of heliophilous shrubs and trees of medium height, such as Rosaceae (Maloideae/Prunoideae) argue for the presence of a rather closed forest formation around Remicourt "En Bia Flo II" at the time of arrival of LBK populations. Differences in wood charcoal assemblages, however,

[6] *Digitaria sanguinalis* (L.) Scop., *D. ischaemum* (Schreb ex Schweigg.) Muhlenb., *Echinochloa crus-galli* (L.) Beauv., *Setaria pumila* (Poir.) Roem et Schult., *S. verticillata* (L.) Beauv., *S.* (L.) Beauv., *S. italica* (L.) Beauv., *Bothriochloa ischaemum* (L.) Keng. There is currently a ninth species present un Belgium, *Panicum miliaceum* L, but its archaeological presence is only attested since the Roman Period onward.

[7] See Annex 1.

Table 2.5 Phytoliths identifications on stone tools from the site of Remicourt "En Bia Flo II"

Attribution	Name ICPN	Further description	#273	#274	#275	#276	#277	#278	#278b	#279inf	#279sup	#280	#281
Nondiagnostic	Elongate triangular section	Or elongate cuneiform		2									
Non diagnostic	Tabular, rectangular, psilate		p	p	p	p	p	p	p	p	p	p	p
Non diagnostic	Cuneiform, carenate, polygonal						1		1				
Non diagnostic	Tabular, rectangular, psilate, with parallel furrows						1						
Poaceae	Elongate clavo-cavate, castellate			1									
Dicot	Tabular, sub-rectangular, polygonal, sinuate,	Puzzle-like, epidermis		2									
Dicot	Tabular, sub-rectangular, polygonal, angular	Puzzle-like but angular			3								
Dicot/Arboreal	Subglobular granulate												1
Hair cell base	Elongated with bulliform end	Drumstick-like											1
UNK70	Subglobular, lateral sides compressed, central internal cavity			1									
UNK8	Parallepipedal, concave, acicular extensions at every corner, and in the middle of 1 length						1	1					1
UNKBL	Cylindric with rings of echinate projection						5						

Table 2.6 Starch grains identifications on stone tools from the site of Remicourt "En Bia Flo II"

Attribution	Observation	273	274	275	276	277	278	278b	279 inf	279 sup	280	281	Total
Cf. *Avena* sp			2										2
Avena sp.	Some are damaged		5	1									6
Triticum sp			11					10			8	1	30
Cf. *Hordeum* sp			6										6
Hordeum sp											64		64
Pisum sp						2							2
Quercus sp	Most are damaged		3	3			2				2		10
Damaged			1										1
Total			28	4		2	2	10			74	1	121

Fig. 2.7 a–d Starch grains from the site of Remicourt "En Bia Flo II": **a** *Avena* granules cluster, ▶ sample #274 under normal and polarized light. **b** Modern *Avena strigosa* Schreb. RBINS reference collection under normal and polarized light. **c** *Hordeum* sp. grain, sample #280 under normal and polarized light. **d** *Hordeum vulgare* L. RBINS reference collection under normal and polarized light. **e–h** Starch grains from the site of Remicourt "En Bia Flo II": **e** cf. *Triticum*, sample #278b under normal and polarized light. **f** Modern *Triticum aestivum* L. RBINS reference collection under normal and polarized light. **g** cf. *Pisum* sp., sample #277 under normal and polarized light. **h** cf. Quercus sp. sample #274 under normal and polarized light

exist between sites in the Hesbaye region, such as an oak dominance at the site of Longchamps or ash tree dominance at the site of *Podrî l'Cortri.*

During the post-pioneer settlement, forest species were still exploited, but together with light-demanding taxa such as Rosaceae, which may indicate an exploitation of the forest edges (Kreuz 1992). The importance of Maloideae is remarkable here (nearly 30% of the sample). Elder, cf. privet, hazel, elm, cf. buckthorn, and maple trees can also be associated with forest edges. The start of a forest opening dynamic in a natural forest progressively favors the availability of light, nutrients, and water for undergrowth plants and explains the increase of species diversity (Ellenberg 1988). Thus, land clearing and forest opening created by pioneer LBK people promoted the development of light-demanding pioneer species such as the Maloideae and Prunoideae subfamily trees in the vicinity of Remicourt, especially near fields and pasture lands. The main limitation of charcoal analysis is the inability to quantify the extent of open areas around LBK settlements. Finally, the riparian forest, which was prone to periodic flooding, was probably not exploited, as indicated by the absence of alder tree and the small proportion of willow/poplar in wood charcoal assemblages. The vegetation changes at Remicourt "En Bia Flo II" fit with the regional dynamic highlighted by charcoal analyses carried out on seven LBK sites in the Geer valley (Salavert et al. 2014) as well as with the present pollen analyses.

Because local edaphic conditions are not different—the ecological setting is the same across the region—and because both the Hesbaye sites are of the same size, with no known subordination relationship between them, HBE cannot be invoked to explain this situation. We definitely think, following HE model, that LBK settlers consciously modified the forest composition for their own purposes across time, and that wood assemblage differences between villages are due to specific cultural traditions, whether symbolic or functional ones, such as the role different taxa may play in sociocultural settings, or the crafting of different objects.

LBK Foodways in Hesbaye

Macroremains analyses indicate a heavy dietary reliance on cereals, mostly wheat with very scarce evidence for barley. Emmer wheat (*T. dicoccum* L.) dominates the assemblage not only at Remicourt "En Bia Flo II" but also at the other sites in

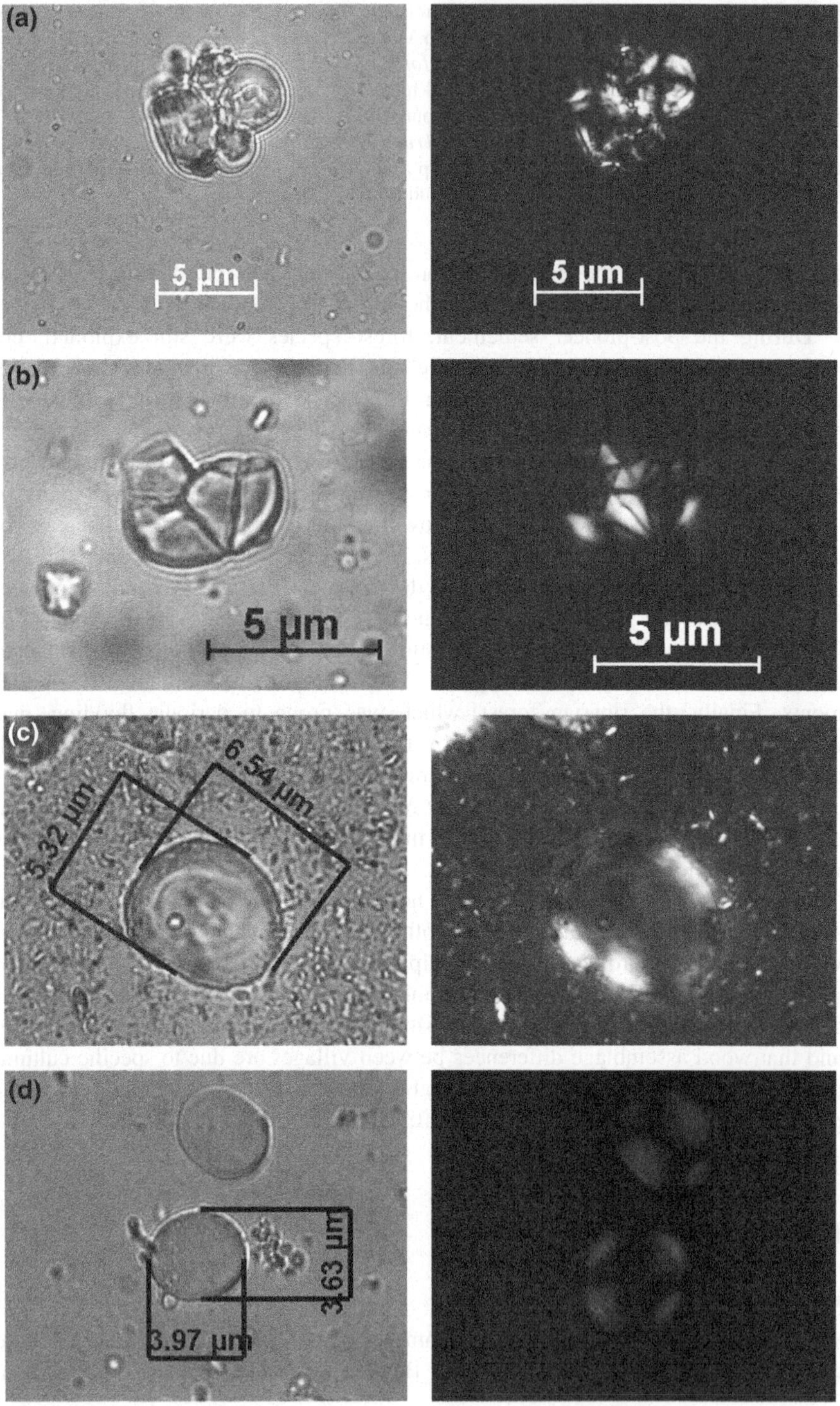
(a)
5 µm
5 µm
(b)
5 µm
5 µm
(c)
6.54 µm
5.32 µm
(d)
3.63 µm
3.97 µm

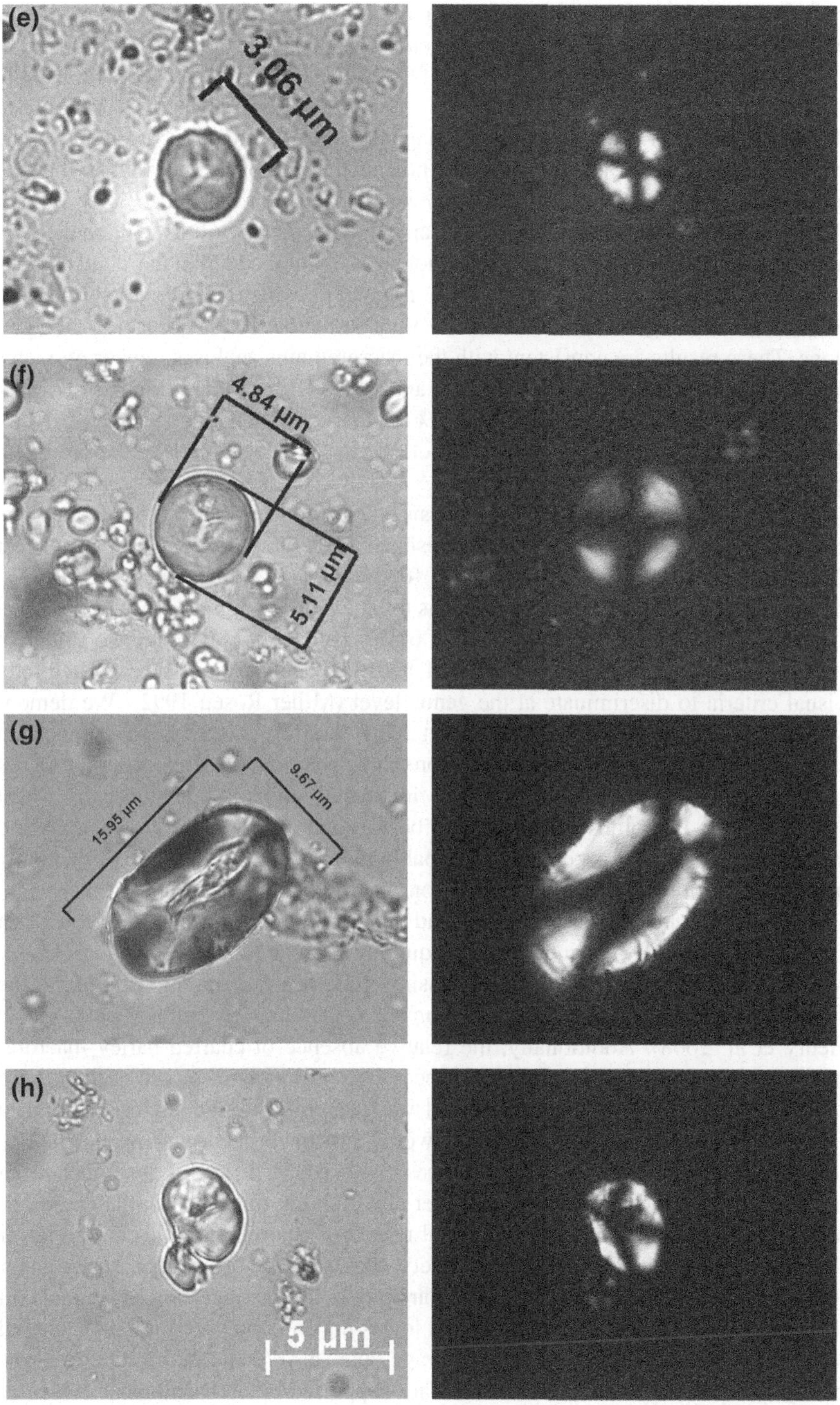

Fig. 2.7 (continued)

central Belgium (Bakels 1992a; Jadin and Heim 2003; Salavert 2011) with few examples of einkorn (*Triticum monococcum* L.).

As we have seen, phytolith analyses of the grindstones at Remicourt "En Bia Flo II" did not provide much information on foodways. Several factors contribute to this relative lack of information. First, diagnostic cereal phytoliths are found in the chaff (glumes, palea, and lemma), but not on the caryopsis themselves that are effectively ground. If cereal grains are well cleaned through successive careful threshing and winnowing, then hardly any diagnostic phytoliths will remain. In addition, long wavy phytolith forms cannot be easily embedded in the small cavities and furrows of the grinding stones, only fragments, which could explain the high number of nondiagnostic forms and the very few related to Poaceae or dicotyledons. These results are consistent with the macroremains analyses: few or no rachis have been identified by Salavert (2010a) and others (Bogaard and Jones 2007: 363; Jones 1984), while chaff constituted 22% of the cereal remains (Salavert 2010a). This means that grains were thoroughly cleaned at the domestic level before being ground locally.

If a very low number of barley caryopsis have been identified, phytolith analyses carried out on sediments from other Hesbaye LBK sites (Chevalier and Bosquet 2010) provide more information about foodways and plant choices (Table 2.7) because of the ability for cereal phytoliths to be identified down to the species level (Ball et al. 1996, 1999; Portillo et al. 2006). In our case, we did not have sufficient counts of phytoliths to permit metric and statistical analyses; thus, we applied only visual criteria to discriminate at the genus level (Miller Rosen 1992). We demonstrated elsewhere (Chevalier and Bosquet 2010) that forms of phytoliths associated with barley (*Hordeum* sp.), whose macroremains were quite scarce, were found not only in half of our contexts, but also in greater number than those of wheat (*Triticum* sp.). Phytoliths of barley have been found in very different contexts and sediments, and thus this species was probably more important in the LBK economy and diet than previously thought based on macroremain data alone.

The starch granules that we have found on the grinding stones at Remicourt "En Bia Flo II" (Table 2.6) confirm the ubiquitous presence of barley in the Hesbay region during the LBK, but also a processing technique that did not use fire, for the starches do not bear any characteristic damage of parching or grilling (Babot 2003; Henry et al. 2008). Additionally, the relative absence of charred barley macroremains can be explained by different processing and cooking practices than those used for wheat. If the barley variety that was present at Remicourt was the naked one, *Hordeum vulgare* var. *nudum*, it would not have needed to be parched to de-hull. This would not have been the case with wheat if parching or other firing processes were used to de-hull the emmer and einkorn grains. Thus, the chance of accidental inclusion in the archaeological record of charred barley would be lower than for wheat. If it was the hulled variety, we can hypothesize that it was either used as fodder, which does not require parching before its ingestion by animals (in this case the six-row barley is best suited for its higher content of proteins), or used to produce some kind of fermented beverage (aka "beer"), and in this case two-row barley was probably used. Neither uses are supported by archaeological evidence,

Table 2.7 Phytolith identifications in soil samples for the sites of the Hesbaye region

Sites	Sample ID#	Concentration	Diversity	Micro wood charcoals	Monocotyledons									
					Poaceae									Cyperaceae
						Pooideae								
					Poaceae	Pooideae nd	*Triticum* sp.	*Hordeum* sp.	*Avena* sp	Panicoideae	Chloridoideae	Bilobate Type 1	Bilobate Type 2	
Fexhe	#02	3	36	2	39	304	3	2		81	58	26		
	#03	0.3	37	0	3	164	19	17		122	43	29		
	#05	1	33	1	35	193	34	14		122	58	1		
	#18	0.3	65	6	23	113	1	1		64	84			
	#10	1.25	27	2		220	22	14	?	86	6	13		
	#07	2.15	27	0	51	207	145	124		61	46			
	#09	1	30	10	17	260	41	156		112	56		10	
	#04	0.5	33	1	42	134	88	25	2	124	54	14		
	#06	0.5	30	2	37	189	12	70		78	65	3	19	
	#17	9	39	8	4	41				36	65	3		
	#08	0.6	24	1	27	172				153	110			
	#16	10	7	6	6	0					5			
	#01	1	34	0	37	206	9	24		89	69	3		
Momalle	#20	0.5	93	12	18	134	24	17	4	19	44			1
	#23	2	34	6	69	79	16	56		30	4			
Remicourt	#24	4	33	3						12	1			
Tombe de Hodeige	#22	1	43	6	15	66	10	17		4	6			
	#21	2	46	7	17	24	3			24	22			
	#11	1	24	7	10	194				115	94			
Voroux/Goreux	#12	1	24	18	20	127				85	125			
	#13	1.15	26	3	10	41				150	89			
	#14	0.75	35	0	35	200				147	44		2	
	#15	7	31	1	63	69		9		137	102		24	

Table 2.7 (continued)

Sites	Dicotyledons				Non diagnostic	Total
	Dicotyledons nd	Arborea	Fruit	Unknown		
Fexhe	5	2		12	603	1135
	10	3		7	41	458
	1			1		459
	36	4		35	60	421
	14	1			232	608
	2			11	106	753
	2			18	229	901
		2			100	585
		7		4	140	624
	7	1		27	122	306
	12				223	697
	12				8	31
	6			4	185	632
Momalle	6		3	110	82	462
	1			12	134	401
Remicourt		13		31	216	273
Hodege's grave	15	4		52	122	311
		6		69	81	246
	7	3			267	690
Voroux/Goreux				27	174	558
	10			21	75	396
	11			2	114	555
	16	5		9	137	571

but the production of alcoholic beverages has been suggested as a trigger for specific plant domestication and cultivation at the Neolithic transition (Wadley and Hayden 2015), and animals clearly needed to be fed during winters.

The presence of starches of oat (*Avena* spp.) on two grinding stones from Remicourt "En Bia Flo II" (Table 2.6) is quite intriguing. Only two macroremains of oat are attested in the Belgian LBK at Weiler-la-Tour/Holzdreisch in Luxembourg (Heim and Jadin 1992), where the specimens were too eroded to allow a determination down to species level. Oat, of any species, is not present anywhere else, as far as we are aware. We do, however, have more evidence from the microfossil record in sediments from nearby sites, Fexhe-le-Haut-Clocher «Podrî l'Cortri» (#4) and Remicourt-Fond de Momalle (#20), where oat phytoliths have been identified (Table 2.7). Since the domesticated oat, *A. sativa* L., appears later in Northwest Europe, around the Late Bronze Age (~1200 BCE), but is cultivated from the LaTène C2 phase (2nd century BCE) onward (Ruas et al. 2012), the oat phytoliths and starches we identified likely come from the wild species, either *Avena fatua* L, *A. sterilis* L.. The latter species appears to have a more southern natural distribution than Belgium, while the first occurrence of *A. strigosa*, Schreb is a macroremain in Northwestern Europe dating from the ninth century CE (Ruas et al. 2012: 349). At the moment we cannot differentiate these different species with starches nor phytoliths. Since this was likely a wild oat, and because of its relative scarcity compared with barley and wheat, we argue that it is a weed from either barley or wheat agricultural fields that was tolerated and eaten together with the other cereals.

Opium poppy seeds (*Papaver somniferum* L.) have been uncovered at Remicourt "En Bia Flo II" in structure 141, associated with the external house (MV), in other words, dating to the earliest LBK phase in Belgium. This is also the case for other sites in the Hesbaye such as Remicourt-Fond de Momalle, Waremme-Vinâve, and Alleur, among others that are not included in this study (Salavert 2010b, 2011). It is unfortunately impossible to determine if the seeds are from the domesticated subspecies *P. somniferum* subsp. *somniferum* or from the wild one *P. somniferum* subsp. *setigerum* (Bakels 1992b; Salavert 2010b); research failed to determine whether opium poppy was cultivated for its oily seeds or for its narcotic properties, or if it was a weed associated with wheat and barley agriculture.

No flax (*Linum usitatissimum* L.) macroremains have been uncovered at Remicourt "En Bia Flo II", but it is present at other sites in the Hesbaye, in particular Remicourt-Fond de Momalle and Alleur.

Macroremain analyses confirm the presence of lentils (*Lens* sp.) as well as edible wild plants, either weeds such as chenopods (*Chenopodium album* L.), or fruits such as hazelnut (*Corylus avellana* L.). Surprisingly, no pips of crab apple (*Malus sylvestris* L.) have been uncovered at Remicourt "En Bia Flo II" though this taxon is actually present as wood charcoal (Maloideae) at Remicourt "En Bia Flo II" and at other sites in the Hesbaye (Salavert 2010b) and elsewhere in Europe (Jacomet 2007; Rottoli 2014). Other ubiquitous taxa including wild prune/cherry, as well as acorn (*Quercus* spp.) or berries such as the blackberry (*Rubus fruticosus* L.), the raspberry (*Rubus idaeus* L.), or the elderberry (*Sambucus* spp.) are also missing from the macroremain record despite systematic sampling, even though their respective wood charcoals have been identified in the Hesbaye region and macroremains are present at other Early Neolithic sites (Jacomet 2007; Rottoli 2014). Acorn and other pulses are, however, present in the microfossil record, for we identified starch grains of *Vicia* sp. and *Quercus* spp. on grinding stones at Remicourt "En Bia Flo II" (Table 2.6).

Agricultural Practices, Territory Exploitation and Cultural Influences

The food plant use at Remicourt "En Bia Flo II" as well as the territorial exploitation of the Hesbaye seem to be quite typical in comparison with other LBK contexts elsewhere in Europe including the appearance of the Near East food plants package and the fast opening of the dense Holocene forest. However, contemporaneous inter-regional differences exist. For instance in Belgium, the Netherlands, and in France, emmer wheat (*T. dicoccum*) clearly dominates assemblages (Bakels 1992a; Jadin and Heim 2003; Salavert 2011), whereas in Germany both emmer wheat (*T. dicoccum*) and einkorn wheat (*T. monococcum*) are present in equal proportion in most of the sites, but not in all sites in particular in Southern Germany, where einkorn dominates assemblages (Kreuz 2007). In the same way, until recently the hulled variety of barley was supposed to be found only on the Eastern bank of the Rhine River, while the naked one was present on the Western side, including Belgium, and Northern France in the Aisne valley (Bakels 1999). However, hulled barley has been found in a LBK site at Alleur, not far from Remicourt (Dietsch-Sellami 1992), and naked barley is present in the Heilbronn–Klinkeberg site in Germany (Stika 1996).

Is there an ecological or physiological reason for this distribution, knowing that einkorn has a much lower yield as compared with emmer, which should therefore be balanced by other gains according to HBE models? The answer is most likely negative and, again, these inter-regional differences are probably rooted in distinctive local cultural traditions (Kreuz 2007; Salavert 2011) as HE purports. Both wheat taxa prefer sunny, dry, and warm weather, as well as middle-alkaline and dry soils, emmer being a little bit more extreme in its ecological requirements (sunnier conditions and more alkaline soils). Einkorn ripens on average one month before emmer, around July. Emmer and einkorn therefore have the same ecological requirements (Le réseau de la botanique francophone 1999). However, einkorn is a more winter-hardy crop and would correspond better to continental conditions found in eastern Germany and Alpine regions, while emmer is better suited to the mild Atlantic winter. If these ecological needs correspond with the archaeological distribution, the physiological characteristics do not fit with it. Indeed, with a same length, between 40 and 80 cm, einkorn stems better resist rainfall than emmer whose stems breaks easily, and would therefore fit better with Northwestern Europe wet summer conditions than emmer when their stems are at their maximum height.

If we look at barley's ecological and physiological characteristics, both the two- and six-row barleys are continental (dry, sunny) plants that like alkaline soils; the 6-row variety tends to be ready for harvesting one month earlier than the 2-row variety, around July. The hulled variety better resists pests and mold in wet summers, which occur in Northwestern Europe, than the naked variety. The archaeological distribution, however, is exactly the opposite: hulled varieties are found more often in continental conditions, in other words in Germany and Alpine regions. To date, and based only on macroremains, barley has been considered related to wheat agriculture as a field weed in a polyculture cereal agriculture (Kreuz 2008; Salavert 2010b). However, the systematic presence of barley microfossils leads us to reconsider this view; more than a weed, barley may have been cultivated together with wheat (maslin) but harvested separately. Barley is indeed slightly bigger than wheat, between 50 cm and 1 m high while wheat height is 40–80 cm; in addition, 6-row barley ripens one month earlier than emmer wheat on average. In this configuration it would have been quite easy to harvest it separately. With other configurations—the presence of 2-row barley and emmer wheat, or 6-row with einkorn wheat—we can hypothesize that specific patches of barley were cultivated as a monoculture. Because only the hulled variety has been found in earliest LBK contexts (Kreuz et al. 2005), this taxon is considered as part of the initial Near Eastern crop package, which is not the case for the naked barley that appears later in LBK II contexts. Based on its distribution in Europe at that time—mostly in a northward corridor from the Mediterranean sea—a Mediterranean cultural influence from the Cardial, Epicardial and Chasséen Neolithics has been proposed to explain its presence in some but not all sites in Northwestern Europe from that period onwards (Salavert 2011).

Opium poppy shows the same inter-regional distribution differences during the LBK II, as well as the same probable Mediterranean origins according to Bakels (1992b) and Salavert (2010a, 2011). As such, at the beginning of the LBK II, opium poppy is most likely considered as a weed associated with barley agriculture, as other weeds also appear at the same time. Weeds are, in fact, of particular interest as we have seen earlier for HE; the weed composition in a given flora is, of course, influenced by the local ecosystem, but as shown by the works of Jones (1984, 1987, 2005; Jones et al. 2000) and Bogaard (2004, 2005; Bogaard et al. 2015), weed composition is also directly linked with agricultural practices. The season of tillage and sowing, the intensity of weed clearings during the growing season, the way harvest is carried on, the instruments used to till, clean, and harvest, as well as fallowing length, field manuring or animal grazing on fields, are of course all influenced by technical and ecological constraints; however, the social organization of economic activities (Van Gijn et al. 2014) and the cultural traditions based on identities (Chevalier 2014) are key to understanding choices made by human groups, and therefore to interpret the remains that we, as archaeobotanists, analyze (Chevalier et al. 2014b). The association of annual weeds that have been found at Remicourt "En Bia Flo II" suggests that agricultural fields were cultivated intensively over long periods (Bogaard 2004, 2005; Kreuz et al. 2005) in small patches without long fallowing, which would allow the growth of perennials and shrubs. This situation means that they practiced manuring, either directly by bringing ruminants to the cereal fields after the harvest, or indirectly by spreading manure before tilling, in order to keep their fields fertile.

Weed composition is also influenced by human travels. Kreuz (2007: 276–281) show in a study carried out on 22 LBK sites that 48 out of 65 weed species were anthropochores, of which 48 have their center of distribution in Southeast Europe. During the LBK II phase, which is the pioneer settlement phase in Belgium, new Mediterranean weed species appear such as *Bromus* cf *arvensis* L., *Bromus sterilis* L., *Malva sylvestris* L., *Galium* cf *verum* L., *Trifolium campestre* Schreb./*dubium* Sibthorp/*arvense* L., and *Vicia hirsuta* (L.) Gray. It is, however, impossible to confirm that the uncovered macroremains in the Hesbaye LBK II sites come from the aforementioned species, for their preservation is not sufficient to allow such a precise identification, and they are idiochores species from the same genus in Middle Belgium. Nevertheless, the presence of theses taxa, among which some species have their natural distribution center in the Mediterranean area, together with opium poppy and barley that have the same probable origin, strong suggest a specific Mediterranean influence in Middle Belgium during the LBK II.

Human Rationality, Cultural History and Social Identities

As shown by Salavert (2010a, 2011), wood charcoal, gathered fruit species and weeds differ both in terms of taxa and of quantities amongst archaeologically contemporaneous Hesbaye settlements, even between settlements that are very close to each other, and therefore in the same ecological conditions. Our own phytolith analyses (Chevalier and Bosquet 2010) brought to light the occurrence of very specific forms in adjacent and potentially contemporaneous LBK settlements (Fig. 2.8).

In addition, differences are attested to on raw materials and object manufacturing shapes. Analyses on ceramic paste carried out at the Field Museum of Natural History Elemental Analysis Facility (Golitko 2010; Golitko et al. 2009) shows that the pattern of association between villages and clay use changes over time: the early peopling phase was associated with ceramics that were made from a Type A2 clay that is similar to Eocene and Oligocene clays, while later occupation ceramics were made also from the Type A1 clay, similar to Cretaceous clays (Bosquet and Golitko 2012). Analyses on flint mineral composition show that grained flints, in particular the Gulpen and Orp types, were predominant in isolated houses, with the pioneer peopling phase, while lithic artifacts made from gray Hesbayen flint were found in the majority of the later occupations but without any change in the *chaine opératoire* or in the kind of objects that were crafted (Bosquet and Golitko 2012).

Finally, differences can also be observed on harvesting tools: in the pioneer house at the site of Waremme-Vinâve only sickle blades of the Karanovo type have been identified, while the Egolzwill type has been identified at Podrî l'Cortri and

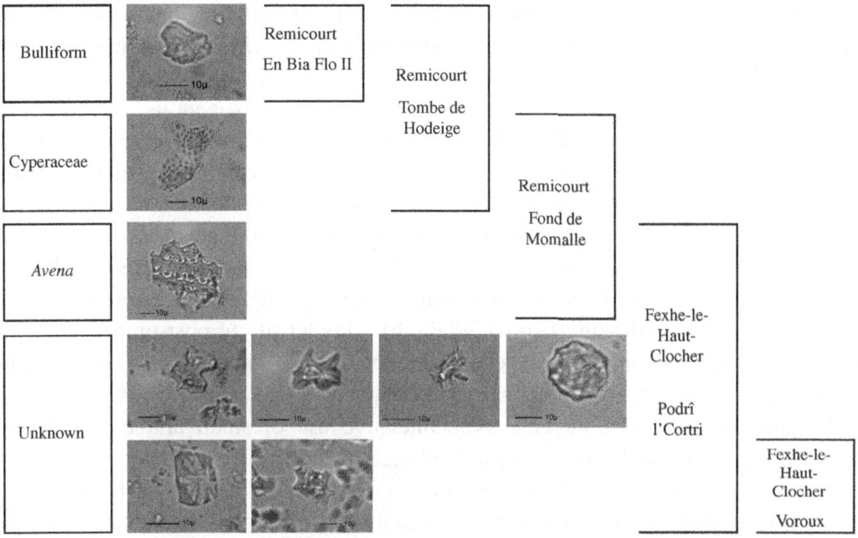

Fig. 2.8 Phytoliths identifications from several sites of the Belgian Hesbaye region (adapted from Chevalier and Bosquet 2010)

Remicourt "En Bia Flo II" post-pioneer settlements, together with the Karanovo type (Beugnier 2005). In fact, the Egolzwill type has been found in greater amounts at those two sites than at other sites such as Darion in Belgium, or even at other sites in Luxemburg or in The Netherlands. In addition, the Egolzwil type sickle blades may show a use-wear specific of a low harvest, close to the soil surface, which could not be identified on the Karanovo type (Beugnier pers. comm.).

Preservation issues, transformation processes that did not use fire or that may have been carried out outside LBK settlements and, above all, the nature of plant remains and the analytical tools used to uncover and analyze the botanical remains may explain the differences observed, both spatially and chronologically. Should we then see these inter- and intraregional differences as mere chronological evolutions? This possibility cannot be excluded for archaeology does not have the temporal resolution to be able to observe changes at the generational level. Indeed, the different post-pioneer settlements of the Hesbaye may have been occupied at slightly different times, imperceptible to us. Nevertheless, to the degree that we can resolve chronology, these settlements appear to have been contemporaneous. So, are there rational causes to explain this diversity of plant exploitation, resource use and tool crafting? Are there specific cultural group trajectories in the Hesbaye that can be read through their remains and their impact on the environment? Or are these differences due to a specific human–nature relationship that we could foresee?

As we have seen, even if there are ecological differences between regions, they do not seem to dramatically influence the choices of plants grown, even if they are less resistant to local conditions or have lower yields. A the regional level, sites do have access to broadly the same vegetation despite their distance—crab apple or hazelnut are quite ubiquitous throughout Belgium for instance, or are so close to each other that they are in the same environment. So these differences are not behavioral responses to specific ecological conditions, and diet breadth cannot be invoked to explain this diversity. Still, they may be due to exogenous constraints, as HBE suggests, such as a demographic increase leading to over-exploit specific resources or patches, forcing LBK settlers to exploit lower ranked resources within the same patch, or to force some part of the population to relocate in another patch with less abundant resources within the given habitat of the Hesbaye.

This scenario has been suggested for the LBK settlements of the Aisne valley in Northern France by Dubouloz (2008), Dubouloz et al. (2012), and elsewhere at the Neolithic transition (Bocquet-Appel 2008a, b). The density of post-pioneer villages in the Hesbaye region, with a very short occupation span (about three to four generations according to the changes of ceramic types and to the C14 dates), would reflect this demographic increase. The intensive use of small arable surfaces in Hesbaye as indicated by the weed-assemblage-provoked acceleration surface erosion and rapid loss of fertility of these surfaces, eventually forcing LBK farmers to move further away every three to four generations to find new fertile arable fields. The rapid change of vegetation at the regional level, as reflected by palynology and wood charcoal analyses, could fit with this overexploitation of resources, for successional vegetation appears everywhere between the pioneer houses and the post-pioneer settlements (Bakels 1992a; Lodewijckx and Bakels 2005; Salavert et al. 2014), in other words after 100 years, which means that the forest cover could not recover completely before a new settlement was built in the same vicinity. At some point, social tensions arising from use rights of limited surfaces of land implied by permanent cultivation (Bogaard 2004: 165), as well as demographic increase would be partially resolved by this model of group splitting and differences observed in plant remains, tools and raw materials, would reflect social constraints on the accessibility of resources. These social tensions may also be reflected by the presence of enclosures surrounding the post-pioneer settlements.

Food practices, however, are determined by a complex ensemble of factors, which makes them relatively insensitive to change. *A fortiori* when changes in the domain of alimentation become perceptible they reflect with certainty profound socioeconomic, symbolic, and political changes (Douglas 1984; Hastorf 1990, 1993; Mead 1997; Mennell 1985; Mintz 1986). We therefore think that HBE models cannot explain the changes occurred between the first settlers and their successors, and the observed diversity during the post-pioneer phase. Above all, we argue that LBK groups in Hesbaye were proactive agents and not simple observers exploiting their environment according to marginal value and opportunity costs and reacting to environmental constraints. LBK settlers definitely created a new landscape in Middle Belgium, shaped the Hesbaye, created a new biodiversity by modifying their environment for agriculture and pasture, and became, in turn, socially and culturally influenced by this new environmental setting, as HE proposes (Balée and Erickson 2006; Ingold 2000). As such, choices in food plants, as well as the technical choices made by cultural groups are also tightly linked to social and identity issues (Chevalier 2014; Chevalier et al. 2014b; Lemonnier

1993). Plant choices as indicated by the different botanical macroremains and microremains, whether used as food, fuel, crafting material or simple contaminants, are as much the result of potential environmental constraints as a necessity to be linked (identified) with a specific part of the landscape, and to differentiate themselves from the "other" through plant choices, which is a common attitude in non-industrial societies (Chevalier et al. 2014b). In the same way, the change of raw material procurement can be understood as a change in the landscape perception and use, not only as a risk benefit issue or a relocation process. The tools to exploit and harvest crops, in particular the ones that constitute the dietary basis such as the cereals in the LBK economy, are not chosen by accident: they must correspond to a specific social and cultural identity. The fact that some LBK villages had different shapes and functionalities of sickle blades (Karanovo vs. Egolswil) should be understood as a social identity statement (Beugnier and Plisson 2004; Rival 1996) and not an attempt to better balance between input and output. Additionally, these different sickle blades reflect different harvesting methods as highlighted by Beunier (pers. comm.) and create different weed assemblages, as observed by Kreuz between the LBK I and IV (2007: 276). Because we lack any direct evidence of violence between LBK settlers, ditches enclosing the post-pioneer settlements cannot be considered only as defensive structures, but probably as some way of restructuring the space, of establishing another new limit between an "inside" and an "outside", from their body to the whole world itself (Descola 2014), since this way of structuring the space was not present during the pioneer phase.

Conclusions

Every kind of archaeological signature constitutes an essential link toward understanding the development and behavior of humankind. Alone, each signature is not meaningful, but put together, and read with the right theoretical lens, they create a history about human groups. This is what we intended to do with the Belgium LBK archaeological contexts we presented here.

We demonstrated that LBK settlers, once established in Hesbaye, specifically chose what they wanted to exploit and use as predicted by HE, with less regard for the location of these sources from their village, contrary to HBE predictions. These cultural choices—specific group's traditions—are expressed in several ways: first in the choice of crop plants, according to both the macro- and microfossil record, whether local ones or "foreign" ones from the Mediterranean cultural sphere; second, in the differences in agricultural tools between villages used to cultivate the same crops; third in weed assemblages that suggest that if LBK people were intensively cultivating very small patches of land, they were cultivating them in different ways; fourth in wood charcoal assemblages which point out to tree species preferences; and finally in raw material selection to craft stone tools and ceramics.

The increase of population density, the multiplication of settlements in the landscape, the modification of the local vegetation, the social tensions arising from land use rights that permanent cultivation implies from the post-pioneer phase onward changed not only the human relationships, but also the human–nature perception, leading to a restructuring of the whole space. The humans created new cultural identities (Tajfel 1982) through differentiated plant use, raw material procurement, and technical solutions. The sense of the natural world changed by the rearranging of environmental elements by the LBK settlers themselves.

It would, however, be illusory to hypothesize how the Hesbaye LBK settlers perceived themselves in the natural world, and were socially and culturally organized through the study of their landscape, as HE suggests (Balée and Erickson 2006). As Descola states (1996, 2005), our 21st century Western ontology, *naturalism*, proceeds by establishing the double dichotomy between nature and culture as well as between human cultures themselves to understand our surrounding world. In fact, HE describes the human relationships with the natural environment as a "collision", to cite Balée and Erickson's own words (2006: 2). According to Descola (1996), this dichotomy is quite recent in the Western world and dates back only from the Renaissance when sciences started to study the natural world as an object per se. Historical Ecology, not to mention Human Behavioral Ecology, may therefore be inadequate theoretical tools if used on non-Western cultural spheres and/or prior the Renaissance period in Europe, for they are based on an inherently distinctive perception of human–nature relationship. Despite this limitation, we argue that Historical Ecology offers the best tools to allow us to grasp non-industrialized or past societies' sociocultural organization and relationship with nature. We should this here with the case of the Hesbaye Early Neolithic settlers, even if further research needs to be carried out to refine the general regional landscape organization, human impact on biodiversity and potential cultural influences from other parts of Europe.

Acknowledgements This contribution was initially a paper presented at the SAA 2011 Fryxell Symposium co-authored by Alexandre Chevalier, Aurélie Salavert, Caroline Hamon, and Mark Golitko. All of these original authors aided in transforming it into a chapter, with insightful comments, and their contributions are evident throughout. We also would like to thank the anonymous reviewers as well as the editors of this book, Maria Bruno and Matt Sayre, for their very constructive comments that helped to improve dramatically our manuscript. But most and foremost, the first author is forever grateful to Christine Hastorf for welcoming him as a post-doctoral researcher for three years at UC Berkeley and initiating him to social archaeology and historical ecology.

Annex 1: Methodologies Used for the Extraction and Identification of Macro- and Microfossils

Wood Charcoals

Layers containing charcoal were systematically sampled during fieldwork in 10 cm arbitrary layers. Soil sample weight was preferred over volume because of field random bloc sampling. Due to their high clay content, sediments were dried at 50 ° C for 24 h prior to their flotation and wet sieving, to allow a better dispersion of soil particles and to avoid any fragmentation. Sediments were then floated and wet sieved using a 250 μm mesh to allow the recovery of all macroremains (Salavert 2010a).

Wood charcoal fragments were split according to their three sections of anatomical observation (transversal, longitudinal tangential, longitudinal radial). The identification of wood charcoal was achieved under a reflection microscope (X50 to X500) with the help of the reference collections available at the Royal Belgian Institute of Natural Sciences and at the French CNRS, UMR 7041 (National Museum of Natural Sciences), as well as with the help of the Schweingruber atlas (1990). Wood charcoals were identified up to the point where the performance curve is flat, in our case 50 fragments after the last new taxon has been identified without new taxon (Salavert 2010a).

Seeds

All the seeds and fruits were identified under magnifying binoculars with the help of the reference collections at the French CNRS, UMR 7041 and at the French Regional Archaeological Centre of Val d'Oise (CRAVO).

Phytoliths

We have applied adapted current published procedures for the phytolith extraction from soils (Madella et al. 1998; Piperno 2006). 5 gr of soils sieved at 500μ were used. Carbonates and organic matted material were removed with HCl (37% concentration) and HNO_3 (69% concentration), respectively. Because of the high clay composition of the analyzed soils, a strong oxydizer was used (1:1 HCl + HNO_3) to remove metallic compounds and break down electric charges. Organic colloids were removed with a 10% solution of KOH for 5 min. The soils were then dispersed with $NaHCO_3$ and continuously stirred up for 12 h. Light soil particles were removed by several one-hour settlings in a 10 cm demineralized water column. Phytoliths were recovered through heavy flotation with [Na6(H2W12O40)H2O], then washed. A standardized quantity of extract was mounted on a microscope slide with Canada Balsam®, covered, and sealed with nail polish. Observation was carried out with a Zeiss microscope under magnification from 125x to 500x. Morphotypes were described according to the ICPN (Madella et al. 2005). A standardized count of 300 short cell was carried out for every microscope slide. Identification was achieved with the help of the RBINS phytolith reference collection, as well as some references that present and describe taxa that may have a relationship with our study area, in particular the Poaceae, arboreal taxa of the temperate zones, as well as herbaceous and arboreal dicotyledons, such as Archer (2009) Fredlund and Tieszen (1994), Fuller (2007), Kaplan et al. (1992), Miller Rosen (1992), Pearsall (2008), Piperno and Pearsall (1998), Twiss et al. (1969). This methodological approach is not unusual for areas that lack extensive reference collections (Neumann et al. 2009; Stromberg 2004).

Starches

For precise procedures regarding stone tool extractions for both phytoliths and starch grains, we refer to Piperno and Holst (1998), Perry (2004), Babot (2005), Piperno (2006), Chandlel-Ezell and Pearsall (2006), and Barton (2007). All the selected grinding stones were thoroughly washed with demineralized water. Subsequently, demineralized water was put on the grinding stones and a sonic toothbrush was used to extract the microfossils from the stone's cavities. The resulting liquid was collected in a beaker and centrifuge to extract the supernatant. A standardized extract was directly mounted with an equal amount of glycerin, covered, and sealed with nail polish. Slides have been completely checked for microfossils under transmission clear and polarized light with a Zeiss microscope under magnification from 125x to 500x.

We have followed Babot's (2005) procedure to describe the starch grains, and used the RBINS reference collection as well as Reichert (1913) for their identification.

All the tools and work surfaces were sterilized and non-powdered glows were worn and changed at all times.

Archer, S. (2009). Williamsburg Phytolith Database. The Colonial Williamsburg Foundation. http://research.history.org/archaeological_research/collections/collarch aeobot/phytolithSearch.cfm. Accessed December 8 2011).

Babot, M. D. P. (2005). *Tecnología y utilización de artefactos de molienda en el Noroeste Prehispánico*. Ph.D., Facultad de Ciencias Naturales e Instituto M. Lillo, Universidad Nacional de Tucumán (San Miguel de Tucumán, Argentina).

Barton, H. (2007). Starch residues on museum artifacts: implications for determining tool use. *Journal of Archaeological Science, 34*, 1752–1762.

Chandler-Ezell, K., & Pearsall, D. (2006). Root and tuber phytoliths and starch grains document manioc (*Manihot esculenta*), arrowroot (*Maranta arundinacea*), and *llerén* (*Calathea* sp.) at the Real Alto site, Ecuador. *Economic Botany, 60*(2), 103–120.

Fredlund, G., & Tieszen, L. (1994). Modern phytolith assemblages from the North American Great Plains. *Journal of Biogeography, 21*(3), 321–335.

Fuller, D. (2007). Archaeobotany: Phytolith teaching and research images. University College London. http://www.homepages.ucl.ac.uk/~tcrndfu/phytoliths.html. Accessed December 8 2011.

Kaplan, L., Smith, M., & Sneddon, L. A. (1992). Cereal grain phytoliths of Southwest Asia and Europe. In G. Rapp & S. Mulholland (Eds.), *Phytolith systematics: emerging issues* (pp. 149–174). New York; London: Plenum.
Madella, M., Alexandre, A., & Ball, T. (2005). International code for phytolith nomenclature 1.0. *Annals of Botany, 96*, 253–260.
Madella, M., Power-Jones, A., & Jones, M. (1998). A simple method of extraction of opal phytolith from sediments using a non-toxic heavy liquid. *Journal of Archaeological Science, 25*, 801–803.
Miller Rosen, A. (1992). Preliminary identification of silica skeletons from Near Eastern archaeological sites: An anatomical approach. In G. Rapp & S. Mulholland (Eds.), *Phytolith systematics* (pp. 129–147). New York; London: Plenun Press. (Advances in Archaeological and Museum Science; 1).
Neumann, K., Fahmyb, A., Lespezc, L., et al. (2009). The Early Holocene palaeoenvironment of Ounjougou (Mali): Phytoliths in a multiproxy context. *Palaeogeography, Palaeoclimatology, Palaeoecology, 276*(1–4), 87–106.
Pearsall, D. (2000). *Paleoethnobotany: A handbook of procedures*. London: Academic Press
Pearsall, D. (2008). Phytoliths in the Flora of Ecuador: The University of Missouri Online Phytolith Database. http://www.missouri.edu/~umcasphyto. Accessed December 8 2011.
Perry, L. (2004). Starch analyses reveal the relationship between tool type and function: An example from the Orinoco valley of Venezuela. *Journal of Archaeological Science, 31*, 1069–1081.
Piperno, D. (2006). *Phytoliths: A comprehensive guide for archaeologists and paleoecologists*. Lanham, MD: Altamira.
Piperno, D., & Holst, I. (1998). The presence of starch grains on prehistoric stone tools from the humid Neotropics: indications of early tuber use and agriculture in Panama. *Journal of Archaeological Science, 25*, 765–776.
Piperno, D., & Pearsall, D. (1998). *The silics bodies of tropical American grasses; morphology, taxonomy, and implications for grass systematics and fossil phytolith identification*. Washington D.C.: Smithsonian Institution Press. (Smithsonian Contribution to Botany; 85).
Reichert, E. T. (1913). *The differentiation and specificity of starches in relation to genera, species, etc*. Washington: Carnegie Instit.
Salavert, A. (2010a). *Apport de l'archéobotanique à la compréhension des sociétés néolithiques. Analyses anthracologiques et carpologiques de neuf sites rubanés de Moyenne Belgique*. Ph.D. thesis, Université de Paris I—Panthéon-Sorbonne, (Paris 2010).
Schweingruber, F. H. (1990). *Anatomie europaïscher Hölzer*—Anatomy of European woods. Eidgenössische Forschungsanstalt für Wald, Schnee und Landschaft, Birmensdorf. Haupt, Bern und Stuttgart.
Stromberg, C. (2004). Using phytolith assemblages to reconstruct the origin and spread of grass-dominated habitats in the great plains of North America during the late Eocene to early Miocene. *Palaeogeography, Palaeoclimatology, Palaeoecology, 207*(3–4), 239–275.

Twiss, P., Suess, E., & Smith, R. M. (1969). Morphological Classification of Grass Phytoliths. *Soil Science Society of America Journal, 33*(1), 109–115.

References

Achar, C., So, J., Agrawal, N., & Duhachek, A. (2016). What we feel and why we buy: The influence of emotions on consumer decision-making. *Current Opinion in Psychology, 10,* 166–170.

Babot, M. D. P. (2003). Starch grain damage as an indicator of food processing. In D. Hart & L. Wallis (Eds.), *Phytolith and Starch Research in the Australia-Pacific-Asian Regions: The state of the art* (pp. 69–81). Canberra: Pandanus Books, Research School of Pacific and Asian Studies, The Australian National University.

Bakels, C. (1992a). The botanical shadow of two early Neolithic Settlements in Belgium: Carbonized seeds and disturbances in a pollen record. *Review of Palaeobotany and Palynology, 73,* 1–19.

Bakels, C. (1992b). Fruits and seeds from the Linearbandkeramik settlement at Meindling, Germany, with special reference to *Papaver somniferum*. *Analecta Praehistorica Leidensiam, 25,* 55–68.

Bakels, C. (1999). Archeobotanical investigations in the Aisne Valley, Northern France, from the Neolithic up to the early Middle Ages. *Vegetation History and Archaeobotany, 8,* 71–77.

Balée, W. (1994). *Footprints of the forest: Ka'apor Ethnobotany—The historical ecology of plant utilization by an Amazonian people*. New York: Columbia University Press.

Balée, W., & Erickson, D. (2006). Time and complexity in historical ecology. In W. Balée & D. Erickson (Eds.), *Time and complexity in historical ecology* (pp. 1–17). New York: Columbia University Press.

Ball, T., Gardner, J., & Anderson, N. (1999). Identifying inflorescence phytoliths from selected species of wheat (*Triticum monococcum, T. dicoccon, T. dicoccoides*, and *T. aestivum*) and barley (*Hordeum vulgare* and *H. Spontaneum*) (Gramineae). *American Journal of Botany, 86* (11), 1615–1623.

Ball, T., Gardner, J., & Brotherson, J. (1996). Identifying phytoliths produced by the inflorescence bracts of three species of wheat (*Triticum monococcum* L., *T. dicoccon* Schrank., and *T. aestivum* L.) using computer-assisted image and statistical analyses. *Journal of Archaeological Science, 23*(4), 619–632.

Beugnier, V. (2005). *Etude techno-fonctionelle de l'industrie lithique du site rubané de Fexhe-le-Haut-Clocher «Podrî l'Cortri»*. Bruxelles: Institut royal des Sciences naturelles de Belgique (unpublished report).

Beugnier, V., & Plisson, H. (2004). Les poignards pressigniens: fonction de signe et fonctions d'usage. In P. Bodu & C. Constantin (Eds.), *Approches fonctionnelles en préhistoire* (pp. 139–154). Paris: Société Préhistorique Française.

Blouet, V., Klag, T., Petitdidier, M. P., & Thomashausen, L. (2013). *Le Néolithique ancien de Lorraine (Mémoires de la société préhistorique française 55)*. Paris: Société Préhistorique française.

Bocquet-Appel, J.-P. (2008a). Explaining the Neolithic demographic transition. In J.-P. Bocquet-Appel & O. Bar-Yosef (Eds.), *The Neolithic demographic transition and its consequences* (pp. 35–56). New York: Springer.

Bocquet-Appel, J.-P. (2008b). The Neolithic demographic transition, population pressure and cultural change. *Comparative Civilizations Review, 58,* 36–49.

Bogaard, A. (2004). *Neolithic farming in Central Europe*. London: Routledge.

Bogaard, A. (2005). Garden Agriculture' and the nature of early farming in Europe and the Near East. *World Archaeology, 35*(2), 177–196.

Bogaard, A., Hodgson, J., Nitsch, E., Jones, G., Styring, A., Diffey, C., et al. (2015). Combining functional weed ecology and crop stable isotope ratios to identify cultivation intensity: A comparison of cereal production regimes in Haute Provence, France and Asturias, Spain. *Vegetation History and Archaeobotany, 25*(1), 57–73.

Bogaard, A., & Jones, G. (2007). Neolithic farming in Britain and Central Europe: Contrast or continuity? *Proceedings of the British Academy, 144,* 357–375.

Boivin, N., Zeder, M., Fuller, D., Crowther, A., Larson, G., Erlandsonh, J., et al. (2016). Ecological consequences of human niche construction: Examining long-term anthropogenic shaping of global species distributions. *PNAS, 113*(23), 6388–6396.

Bosquet, D., Fock, H., Goffioul, C., Preud'Homme, D., & Marchal, J.-P. (2004). La néolithisation en Hesbaye (Belgique): Apport des fouilles récentes (1996–1999) sur le tracé du TGV oriental et le domaine militaire d'Alleur. In *Section 9 et 10. Néolithique et âge du cuivre* (pp. 81–94, BAR Int. Series, S1303). Oxford: Archaeopress.

Bosquet, D., & Golitko, M. (2012). Highlighting and characterizing the pioneer phase of the Hesbayan LBK (Liege Province, Belgium). In S. Wolfram, H. Stauble, M. Cladders, & Th. Tischendorf (Eds.), *Siedlungsstruktur und Kulturwandel in der Bandkeramik, Beitrage der Internationale Tagung "Neue Fragen zur Bandkeramik oder Alles beim Alten ?!", Leipzig, 23–24. September 2010* (pp. 91–106, Arbeits: und Forschungsberichte zur sächischen Bodendenkmalpflege 25). Dresden: Landesmat für Archäologie.

Bosquet, D., Golitko, M., & Salavert, A. (2008). Une phase pionnière à l'origine du peuplement rubané de la Hesbaye liégeoise (Belgique). In M. Ilett, L. Burnez Lanott, & P. Allard (Eds.), *Fin des traditions danubiennes dans le Néolithique du Bassin parisien et de la Belgique (5100–4700 BC)* (pp. 301–315, Mémoires de la Société préhistorique française XLIV). Paris: Société préhistorique française.

Bosquet, D., & Preud'Homme, D. (1998). Dernière campagne de fouille sur le village rubané de Remicourt au lieu dit "En Bia Flo" II. *Notae Praehistoricae, 18,* 119–122.

Cahen, D., Keeley, L. H., & Jadin, I. (1990). Trois villages fortifiés du Rubané récent en Hesbaye liégeoise. In D. Cahen, & M. Otte (Eds.), *Rubané et Cardial* (pp. 125–146, Études et Recherches Archéologiques de l'Université de Liège 39). Liège: Université de Liège.

Chabal, L. (1994). Apports récents de l'anthracologie à la connaissance des paysages passés: performances et limites. *Histoire & Mesure, 9*(3/4), 317–338.

Chevalier, A. (2014). Social status, identity and contexts: Introduction. In A. Chevalier, E. Marinova, & L. Peña-Chocarro (Eds.), *Plants and people: Choices and diversity through time* (pp. 385–390). Oxford: Oxbow.

Chevalier, A., & Bosquet, D. (2010). Apports de l'analyse des phytolithes en contexte détritique Néolithique ancien: résultats obtenus sur cinq sites rubanés de Hesbaye liégeoise (Belgique). In C. Delhon, I. Théry-Parisot, & S. Thiébault (Eds.), *Des hommes et des plantes. Exploitation du milieu et gestion des ressources végétales de la préhistoire à nos jours* (pp. 33–68). Antibes: Eds ADPCA.

Chevalier, A., & Bosquet, D. (2013). Culture matérielle, exploitation du territoire et identités socio-culturelle rubanées en Belgique: analyses de microfossiles sur des instruments de mouture. In P. Anderson, C. Cheval, & A. Durand (Eds.), *an interdisciplinary focus on plant-working tools* (pp. 189–204). Antibes: Eds ADPCA.

Chevalier, A., Marinova, E., & Peña-Chocarro, L. (2014a). Factors and issues in plant choices. In A. Chevalier, E. Marinova, & L. Peña-Chocarro (Eds.), *Plants and people: Choices and diversity through time* (pp. 3–13). Oxford: Oxbow.

Chevalier, A., Marinova, E., & Peña-Chocarro, L. (Eds.). (2014b). *Plants and people: Choices and diversity through time*. Oxford: Oxbow.

Constantin, C., & Demarez, L. (1984). Cinq années de fouilles dans le groupe de Blicquy (Belgique). *Revue archéologique de Picardie, 1–2,* 73–86.

Descola, P. (1996). Constructing natures: Symbolic ecology and social practices. In P. Descola & G. Pálsson (Eds.), *Nature and society: Anthropological perspectives* (pp. 82–102). New York: Routledge.

Descola, P. (2005). *Par-delà nature et culture*. Paris: Gallimard.

Descola, P. (2014). Anthropologie de la nature. *Annuaire du Collège de France 2012–2013*. Résumé des cours et travaux 113e année, pp. 679–701.

Dietsch-Sellami, M.-F. (1992). Carpologie de la fosse 4.26 (campagne 2000) du site néolithique ancient " Le Domaine militaire " à Alleur (Province de Liège, Belgique). *Notae Praehistoricae (Belgium), 22*, 109–115.

Douglas, M. (1984). *Food in the social order: Studies of food and festivities in three American communities*. New York: Russell Sage Foundation.

Dubouloz, J. (2008). Impacts of the Neolithic demographic transition on linear pottery culture settlement. In J.-P. Bocquet-Appel & O. Bar-Yosef (Eds.), *The Neolithic demographic transition and its consequences* (pp. 207–235). New York: Springer.

Dubouloz, J., Chartier, M., Hachem, L., & Ilett, M. (2012). «Interdépendance et cohésion des différents niveaux de territorialité au Néolithique rubané en Bassin parisien». In V. Carpentier & C. Marcigny (Eds.), *Des Hommes aux Champs. Pour une archéologie des espaces ruraux du Néolithique au Moyen Âge* (pp. 23–34). Rennes: Presses universitaires de Rennes.

Ellenberg, H. (1988). *Vegetation ecology of Central Europe*. Cambridge: Cambridge University Press.

Foxall, G. (2009). *Interpreting consumer choice: The behavioral perspective model*. London: Routledge.

Gallay, A. (1995). A propos des travaux récents sur la Néolithisation de l'Europe de l'Ouest. In J.-L. Voruz (Ed.), *Chronologies néolithiques* (pp. 17–25). Ambérieux-en-Bugey: Société Préhistorique Rhodanienne.

Golitko, M. (2010). *Warfare and Alliance Building during the Belgian Early Neolithic, late sixth Millennium BC*. Ph.D. Thesis, University of Illinois, Chicago.

Golitko, M. (2015). *LBK Realpolitik : An Archaeometric study of conflict and social structure in the Belgian early Neolithic*. Oxford: Archaeopress.

Golitko, M., Bosquet, D., Keeley, L., Jadin, I., & Goemaere, E. (2009). *LA-ICP-MS Analyses of Early Neolithic (LBK Culture) Ceramics and Clay Samples from the Hesbaye region of Belgium*. Poster presented at the 73rd annual meeting of the Society for American Archeologist, Atlanta, Georgia, 22–26 April.

Hastorf, C. A. (1990). The effect of the Inka state on Sausa agricultural production and crop consumption. *American Antiquity, 55*, 262–290.

Hastorf, C. A. (1993). *Agriculture and the onset of political inequality before the Inka*. Cambridge: Cambridge University Press.

Hastorf, C. A. (2003). Andean luxury foods: Special food for the ancestors, deities and the élite. *Antiquity, 77*(29), 545–554.

Heim, J. (1985). Recheches sur l'environnement paléoboatnique du vallage rubané de Darion par l'étude des pollens et de restes de diaspores (graines). *Bulletin de la Société royale belge d'Anthropologie et de Préhistoire, 96*, 31–48.

Heim, J., & Jadin, I. (1992). Paléobotanique des sites rubanés et Weiler-La-Tour: Holzdreisch et Alzingen-Grossfeld. *Bulletin de la Société Préhistorique Luxembourgeoise, 13*, 37–58.

Henry, A., Hudson, H., & Piperno, D. (2008). Changes in starch grain morphologies from cooking. *Journal of Archaeological Science, 36*(3), 915–922.

Ingold, T. (1996). The optimal forager and economic man. In P. Descola & G. Pálsson (Eds.), *Nature and society: Anthropological perspectives* (pp. 25–44). New York: Routledge.

Ingold, T. (2000). *The perception of the environment: Essays on livelihood, dwelling and skill*. London: Routledge.

Jacomet, S. (2007). Neolithic plant economies in the Northern Alpine Foreland from 5500–3500 cal BC. In S. Colledge & J. Connolly (Eds.), *The origins and spread of domestic plants in Southwest Asia and Europe* (pp. 221–258). Walnut Creek, CA: Leftcoast Press.

Jadin, I. (2003). *Trois petits tours et puis s'en vont... La fin de la présence danubienne en Moyenne Belgique*. Liège: Université de Liège.

Jadin, I., & Heim, J. (2003). Sur la voie de l'orge et du pavot : macrorestes végétaux et agriculture rubanée du Haut Geer dans un cadre européen. In I. Jadin (Ed.), *Trois petites tours et puis s'en vont... La fin de la présence danubienne en Moyenne-Belgique* (pp. 345–392). Liège: Université de Liège.

Jones, G. (1984). Interpretation of archaeological plant remains: Ethnographic models From Greece. In W. Van Zeist & W. A. Casparie (Eds.), *Plants and ancient man* (pp. 43–61). Rotterdam: Balkema.

Jones, G. (1987). A statistical approach to the archaeological identification of crop processing. *Journal of Archaeological Science, 14,* 311–323.

Jones, G. (2005). Garden cultivation of staple crops and its implications for settlement location and continuity. *World Archaeology, 37*(2), 164–176.

Jones, G., Bogaard, A., Charles, M., & Hodgson, J. (2000). Distinguishing the effects on agricultural practices relating to fertility and disturbance: A functional ecological approach in archaeobotany. *Journal of Archaeological Science, 27*(11), 1073–1084.

Keeley, L., Bosquet, D., Quick, R., Jadin, I., & Golitko, M. (2005). Nouvelles fouilles sur le site rubané de Waremme-Longchamps. Rapport de la campagne 2005. *Notae Praehistoricae, 25,* 153–161.

Kennet, D., & Winterhalder, B. (2006). *Behavioral ecology and the transition from hunting and gathering to culture.* Berkeley and Los Angeles: University of California Press.

Kreuz, A. (1992). Charcoal from ten early Neolithic settlements in Central Europe and its interpretation in terms of Woodland Management. In J.-L. Vernet (Ed.), *Les charbons de bois, les anciens écosystèmes et le rôle de l'homme* (pp. 383–394). Paris: Société botanique de France.

Kreuz, A. (2007). Archaeobotanical perspectives on the beginning of agriculture North of the Alps. In S. Colledge & J. Connolly (Eds.), *The origins and spread of domestic plants in Southwest Asia and Europe* (pp. 259–294). Walnut Creek, CA: Leftcoast Press.

Kreuz, A. (2008). Closed forest or open woodland as natural vegetation in the surroundings of Linearbandkeramik Settlements? *Vegetation History and Archaeobotany, 17,* 51–64.

Kreuz, A., Marinova, E., Schäfer, E., & Wiethold, J. (2005). A comparison of early Neolithic crop and weed assemblages from the Linearbandkeramik and the Bulgarian Neolithic cultures: Differences and similarities. *Vegetation History and Archaeobotany, 14,* 237–258.

Le réseau de la botanique francophone (1999). Association Tela Botanica. http://www.tela-botanica.org. Accessed 23 June 2014.

Lemonnier, P. (1993). *Technological choices: Transformation in material cultures since the Neolithic.* New York: Routledge.

Livingstone-Smith, A., Beugnier, V., Bosquet, D., Collette, O., Defgnée, A., Deligne, F., et al. (2012). Habitats du Néolithique ancien en Hainaut occidental (Ath et Beloeil, Belgique): Ormeignies «Le Pilori» et Aubechies «Coron Maton». *Etudes et Documents, Archéologie, Namur, 18,* 0–278.

Lodewijckx, M. (1990). *Les deux sites rubanés de Wange et Overhespen (Belgique, prov. Brabant)* (Rubané et Cardial). Liège: Université de Liège.

Lodewijckx, M., & Bakels, C. (2005). The interaction between early farmers and indigenous people in Central Belgium. http://millennium.arts.kuleuven.ac.be/Wea/Farmers/index.htm. Accessed 16 July 2016.

Mason, R. (1998). *The economics of conspicuous consumption: Theory and thought since 1700.* Cheltenham, UK: Edward Elgar.

Mazurié de Keroualin, K. (2001). *Genèse et diffusion de l'agriculture en Europe.* Paris: Errance.

Mead, M. (1997). The changing significance of food. In C. Counihan & P. Van Esterik (Eds.), *Food and culture: A reader* (pp. 11–19). New York: Routledge.

Meier-Arendt, W. (1966). *Die bandkeramische Kultur im Untermaingebiet.* Amtes für Bodendenkmalpflege im Regierungsbezirk Darmstadt, Habelt in Kommission, Bonn.

Mennell, S. (1985). *All manners of food: Eating and taste in England and France from the Middle Ages to the present.* Oxford: Blackwell.

Miller Rosen, A. (1992). Preliminary identification of silica skeletons from Near Eastern Archaeological sites: An anatomical approach. In G. Rapp & S. Mullholland (Eds.), *Phytolith systematics* (pp. 129–147, Advances in Archaeological and Museum Science 1). London: Plenum Press.

Mintz, S. (1986). *Sweetness and power: The place of sugar in modern history*. New York: Penguin.
Modderman, P. J. R. (Ed.). (1970). *Linearbandkeramik aus Elsloo und Stein* (Analecta Praehistorica Leidensia 3). 's-Gravenhage: Staatsverlag.
Portillo, M., Ball, T., & Manwaring, J. (2006). Morphometric Analysis of Inflorescence Phytoliths Produced by *Avena sativa* L. and *Avena strigosa* Schreb. *Economic Botany, 60*(2), 121–129.
Rival, L. (1996). Blowpipes and spears: The social significance of Huaorani technological choices. In P. Descola & G. Pálsson (Eds.), *Nature and society: Anthropological perspectives* (pp. 145–164). London: Routledge.
Roosens, H. (1961). Bandkeramische nederzetting. *Archéologie, 1*, 519–520.
Rottoli, M. (2014). Before the empire: Prehistoric fruit gathering and cultivation in Northern Italy. In A. Chevalier, E. Marinova, & L. Peña-Chocarro (Eds.), *Plants and people: Choices and diversity through time* (pp. 163–169). Oxford: Oxbow.
Ruas, M.-P., Zech-Matterne, V., Dietsch-Sellami, M.-F., Pradat, B., & Preiss, S. (2012). Les avoines dans les productions agro-pastorales du Nord-Ouest de la France. Données carpologiques et indications textuelles. In V. Carpentier & C. Marcigny (Eds.), *Des Hommes aux Champs: pour une archéologie des espaces ruraux du Néolithique au Moyen-Age* (pp. 327–365). Rennes: Presses Universitaires de Rennes.
Salavert, A. (2010a). *Apport de l'archéobotanique à la compréhension des sociétés néolithiques. Analyses anthracologiques et carpologiques de neuf sites rubanés de Moyenne Belgique*. Ph.D Thesis, Université de Paris I—Panthéon-Sorbonne, Paris.
Salavert, A. (2010b). Le pavot (*Papaver somniferum*) à la fin du 6e millénaire av. J.-C. en Europe occidentale. *Anthropobotanica, 1*(3), 3–15.
Salavert, A. (2011). Plant economy of the first farmers of Central Belgium (Linearbandkeramik, 5200–5000 B.C.). *Vegetation History and Archaeobotany, 20*(5), 321–332.
Salavert, A., Bosquet, D., & Damblon, F. (2014). Natural woodland composition and vegetation dynamic during the Linearbandkeramik in North-Western Europe (Central Belgium, 5200–5000 B.C.). *Journal of Archaeological Sciences, 51*(1), 84–93.
Smith, B. (2012). A cultural niche construction theory of initial domestication. *Theoretical Biology, 6*, 260–271.
Smith, B. (2015). A comparison of niche construction theory and diet breadth models as explanatory frameworks for the initial domestication of plants and animals. *Journal of Archaeological Research, 23*(3), 215–262.
Smith, E., & Winterhalder, B. (Eds.). (1992). *Evolutionary ecology and human behavior*. New York: Aldeline de Gruyter.
Stika, H.-P. (1996). *Vorgeschichtliche Pflanzenreste aus Heilbronn-Klingenberg. Archäobotanische Untersuchungen zum Michelsberger Erdwerk auf dem Schlossberg* (Bandkeramik, Michelsberger Kultur, Späthallstatt/Frühlatene). Stuttgart: Konrad Theiss Verlag.
Tajfel, H. (1982). *Social identity and intergroup relations*. Cambridge: Cambridge University Press.
Van Gijn, A., Whittaker, J., & Anderson, P. (Eds.). (2014). *Exploring and explaining diversity in agricultural technology*. Oxford: Oxbow.
Wadley, G., & Hayden, B. (2015). Pharmacological Influences on the Neolithic Transition. *Journal of Ethnobiology, 35*(3), 566–584.
Winterhalder, B., & Kennett, D. (2006). Behavioral ecology and the transition from hunting and gathering to agriculture. In D. Kennett & B. Winterhalder (Eds.), *Behavioral ecology and the transition from hunting and gathering to agriculture* (pp. 1–21). Berkeley, Los Angeles: University of California Press.
Zeder, M. (2012). The broad spectrum revolution at 40: Resource diversity, intensification, and an alternative to optimal foraging explanations. *Journal of Anthropological Archaeology, 31*, 241–264.
Zohary, D., Hopf, M., & Weiss, E. (2012). *Domestication of plants in the old world: The origin and spread of cultivated plants in West Asia, Europe, and the Nile Valley*. Oxfrod: Clarendon Press.

Chapter 3
Cultigen Chenopods in the Americas: A Hemispherical Perspective

Gayle J. Fritz, Maria C. Bruno, BrieAnna S. Langlie, Bruce D. Smith and Logan Kistler

Introduction

Few if any of us working with archaeological plant remains 30 years ago dreamed that a chenopod could by now have achieved Supergrain status in the popular food world. Back then, North American chenopod was considered a lowly weed by most archaeologists, and quinoa was not well known outside of Peru and Bolivia. Now, of course, quinoa is the darling of celebrity chefs around the world, even featured on the cover of *Time* Magazine's September 1, 2011 issue. A Google search for quinoa recipes will turn up millions of results. Boxes of quinoa are sold in chain supermarkets across North America, it is available in bulk at stores catering to the health-conscious, entire cookbooks are dedicated to this single ingredient, and quinoa dishes are offered several times a week in the main cafeterias at our colleges

G.J. Fritz (✉)
Department of Anthropology, Washington University, St. Louis, USA
e-mail: gjfritz@wustl.edu

M.C. Bruno
Department of Anthropology and Archaeology, Dickinson College, Carlisle, USA
e-mail: brunom@dickinson.edu

B.S. Langlie
Department of Anthropology, Loyola University Chicago, Chicago, USA
e-mail: blanglie@luc.edu

B.D. Smith
Program in Human Ecology and Archaeobiology and Department of Anthropology, National Museum of Natural History, Smithsonian Institution, Washington, D.C., USA
e-mail: smithb@si.edu

L. Kistler
Department of Anthropology, National Museum of Natural History, Smithsonian Institution, MRC 112, 10th and Constitution Ave NW, Washington, D.C., USA
e-mail: KistlerL@si.edu

M.P. Sayre and M.C. Bruno (eds.), *Social Perspectives on Ancient Lives from Paleoethnobotanical Data*, DOI 10.1007/978-3-319-52849-6_3

and universities. Its virtues are widely appreciated: a subtle nutty flavor, gluten-free, fuller suite of amino acids, and higher protein content than staple cereals such as rice, wheat, or maize (National Research Council 1989). UNESCO (General Assembly Resolution 66/221) declared 2013 the International Year of Quinoa in recognition of its high nutritional value, deep cultural roots in the Andes, and its potential to aid in resolving world hunger.

This meteoric rise in public appreciation is paralleled by archaeological research that magnifies our understanding of the role and significance played by members of the genus *Chenopodium* in the pre-Colonial Americas. In 1980, quinoa was of course recognized as an important Andean crop, but the timing of its domestication and geographic particulars of its production were not well understood. In eastern North America, *Chenopodium* was included in the pre-maize Eastern Agricultural Complex (EAC), but its domesticated (vs. wild or weedy) status was questioned, and it seemed reasonable to suppose that if domesticated chenopod had been grown in eastern North America, it was probably introduced from Mesoamerica (Wilson 1981). Phylogenetic relationships among species in North, Central, and South America were of considerable concern (Hunziker 1952; Wahl 1952; Wilson 1990; Wilson and Heiser 1979), but the miracles of modern molecular biology were in their infancy, and ancient DNA research was unborn.

What has been learned about *Chenopodium* since, and what we focus on in this chapter, extends far beyond the initial question of how to recognize domesticated *versus* wild populations in the archaeological record. Here, we summarize recent contributions made by archaeologists and colleagues from other disciplines toward understanding the many factors involved in the domestication of *Chenopodium* in North and South America (Fig. 3.1). We focus on these two regions and the domesticated forms of *C. berlandieri* and *C. quinoa*, for two reasons. First, advancements in both regions have emerged from mentorships and collaborations of researchers in these two areas. Second, both of these crops appear to have played a central role in the development of complex societies in each region, yet their roles took very separate paths. Thus, a comparison of chenopod research in North and South America sheds light on processes of domestication and intensification of this particular crop, and also contributes to broader discussions of agricultural developments worldwide.

We begin with research that provides the intellectual and collaborative link between the two regions: the study of seed morphology. While new methods to document micro-morphological markers of domestication in chenopod were first employed in eastern North America, use of these techniques has subsequently clarified what was a complicated process of quinoa domestication in South America and pointed toward a greater diversity than initially anticipated. Since the establishment of the domesticated status of North American chenopods, morphological traits have been used to identify distinct subspecies or varieties of this crop. We then turn to advances in molecular studies, which have recently clarified the independent domestication of chenopods in several different regions across the Americas, but particularly in North America. Molecular studies in the Andes have

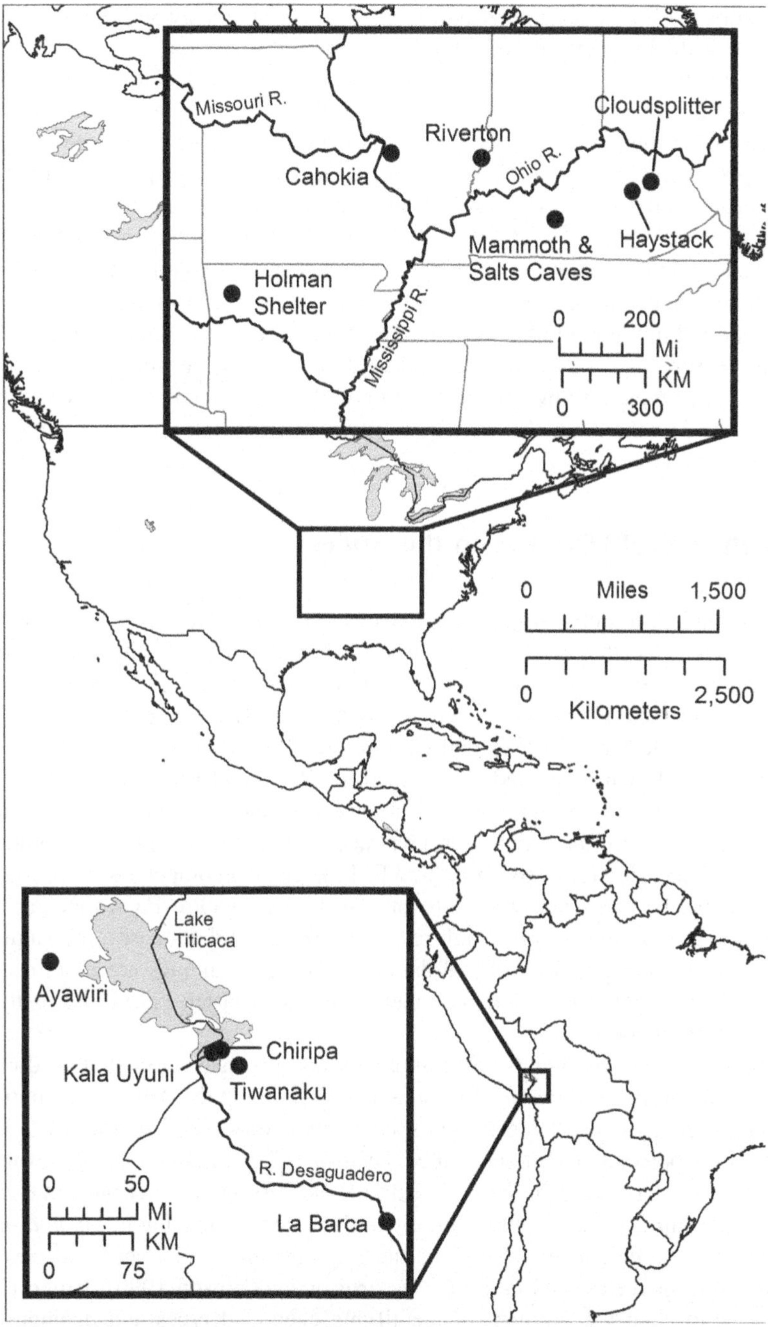

Fig. 3.1 Location of sites mentioned in text

focused primarily on agronomic issues, but many scholars are still working toward identifying the progenitor(s) of quinoa.

These advances in studying the morphology and molecular components of chenopod domestication permit us to explore the more pressing questions of why members of the genus became crops in both regions and what role(s) they played in later agricultural systems of these respective areas. These crops constituted food that was central to and inseparable from considerations of identity, status, ritual, exchange, and sociopolitical life in both of these regions. A comparison of the social and political contexts of chenopod domestication is significant because while this human intervention resulted in very similar biological consequences for the plant, the cultural conditions and consequences for domestication and later intensification were quite different. An appreciation of these chenopods as important foods in each region allows us to reflect upon their individual trajectories in North America and in the Andes.

Morphological Diversity in the Andes

The foundation of archaeological approaches to *Chenopodium* domestication is the study of morphometric attributes of the seeds themselves. Collecting measurements on attributes such as seed size and seed coat thickness using both light and scanning electron microscopy was pioneered by Wilson (1981), Smith (1985a, b), Fritz (1986), Fritz and Smith (1988), and Gremillion (1993) for eastern North American chenopods and then employed by Nordstrom (1990) and Eisentraut (1998) in the Andes. New digital technologies for microscopy have enhanced our ability to collect this type of information with great accuracy and speed. Seed coat thickness can now be measured directly from the SEM image on a computer screen, and seed diameter can be calculated with software (such as ImagePro Plus) that traces the circumference of the seed on a high-resolution image. In both cases, the values are automatically entered into a digital database. These techniques are more accurate than measuring with an ocular micrometer and much faster, permitting us to take more measurements.

Recent research highlights the complexity and diversity of chenopod domestication in the Andean *altiplano*. The botanical situation in the Andes is complicated by the presence of two domesticated species: the well-known quinoa (*C. quinoa*); and a less-appreciated cultigen called kañawa (*C. pallidicaule*), still grown by traditional farmers, often in the most extreme and risk-prone environmental zones. Presenting further challenges, quinoa has a weedy companion, *quinoa negra* (*C. quinoa* var. *melanospermum*), and there is a common wild/weedy species with medicinal properties called *paiko* (*C. ambrosioides*) (Wilson 1990). Bruno (2001, 2006) approached the problem of identifying chenopod types and distinguishing between domesticates and their wild or weedy relatives by coding for a combination of quantitative and qualitative attributes: seed coat thickness; seed diameter; ratio of seed coat thickness to diameter; margin configuration; and seed coat texture.

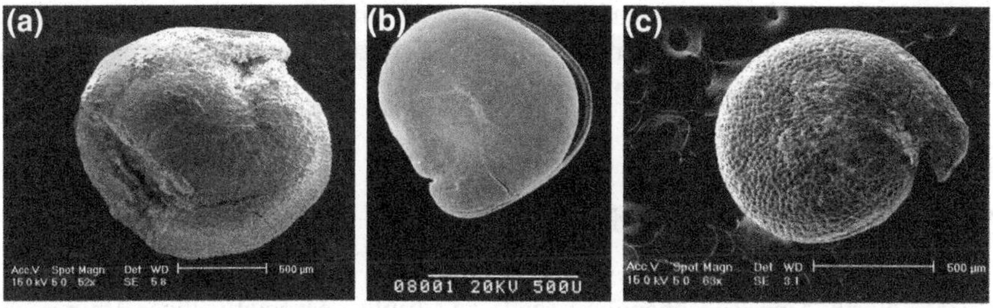

Fig. 3.2 **a** Scanning electron microscope image of a carbonized domesticated quinoa seed (covered by pericarp) from Chiripa, Bolivia, specimen #CCH89. **b** Scanning electron microscope of La Barca Type 1 specimen #LB108. **c** Scanning electron microscope image of a wild kañawa type (Unknown Amaranthaceae Bruno 2008; La Barca Type 2, Langlie et al. 2011) from Chiripa, Bolivia specimen #WU8001. All scales are 500 microns (0.5 mm)

This approach enabled Bruno and Whitehead (2003) to document the economic importance of domesticated quinoa and the existence of a crop–weed complex from as early as 1500 B.C.. at Chiripa on the Taraco peninsula in the southwest Lake Titicaca Basin. Recently, Bruno has been working on identifying a chenopod type from the Taraco samples with seeds that are smaller in diameter than quinoa or quinoa negra, have a smooth to canaliculated seed coat, and are round in shape, which she labeled as "unknown Amaranthanceae" (Bruno 2008, pp. 292–296) (Fig. 3.2). Analysis is ongoing for comparative collections from Bolivia of wild kañawa seeds, and it is likely that they are a wild form of kañawa (Bruno et al. 2013).

To the south, in the Bolivian province of Oruro, BrieAnna Langlie and colleagues (2011) described a previously unrecognized morphological type of thin-testa chenopod from the La Barca site, a Formative (1500-400 B.C.) Wankarani complex village. After reading Gordon's (2006) work on modern Mexican chenopods, Langlie added the attribute of "beak prominence" to Bruno's previous set of seed characteristics, and she used a new digital technique for standardized measurement of seed diameter (Fig. 3.2). Food producers at La Barca grew a distinct type of *Chenopodium* (referred to as La Barca Type 1) that does not correspond to known, modern varieties. It may have eventually crossbred with other domesticated species or varieties, or, alternatively, have gone extinct. Langlie et al. (2011) also recognized a second, less common seed type at La Barca (Type 2) that has similar attributes to the possible wild kañawa described by Bruno from the Taraco Peninsula.

With increased use of flotation and other fine-grained recovery methods from sites throughout the South American Andes, chenopods are frequently being recovered and the details of their morphology are being examined (López et al. 2015). Archaeobotanists are using these measures to describe the morphological traits of seeds to determine their domesticated status (Planella et al. 2010, 2011) as well as describe their diversity in later periods (López and Nielsen 2012). This work not only contributes to the understanding of when and where particular crops were grown (Planella et al. 2014), but it opens our minds to patterns of selection that can vary across ecological zones, and it raises the potential of exploring seed exchange, cultural interaction, and the formation of regional cuisines. With this potential in mind, we now return to eastern North America.

New Morphological Evidence in Eastern North America

Eastern North American chenopod seeds and fruits—especially those from rock-shelter caches—were pivotal in early SEM studies where researchers established baseline methods for measuring testa thickness and scrutinizing coat texture of directly AMS-dated specimens (Smith 1985a, b; Wilson 1981). The Flotation Revolution opened the sluiceways to recovery of literally millions of chenopod seeds from storage pits, trash pits, and other contexts across the Midwest, Upland South, and Trans-Mississippi South. Thin-testa cultigen forms had been bred by 1800 B.C., and farmers grew chenopod along with other members of the EAC throughout the first millennium B.C., the first millennium A.D., and into the first half of the second millennium A.D. (Fig. 3.3). The domestication process, as reflected in a reduction in seed coat thickness, resulted from automatic seedbed competition (DeWet and Harlan 1975), in conjunction with intentional selection by early cultivators.

Newly excavated material has broadened our views, but so has reanalysis of samples uncovered decades ago. In 2009, Smith and Yarnell published an article showing that both pale and dark, thin-testa chenopod seeds dating to 1800 B.C.

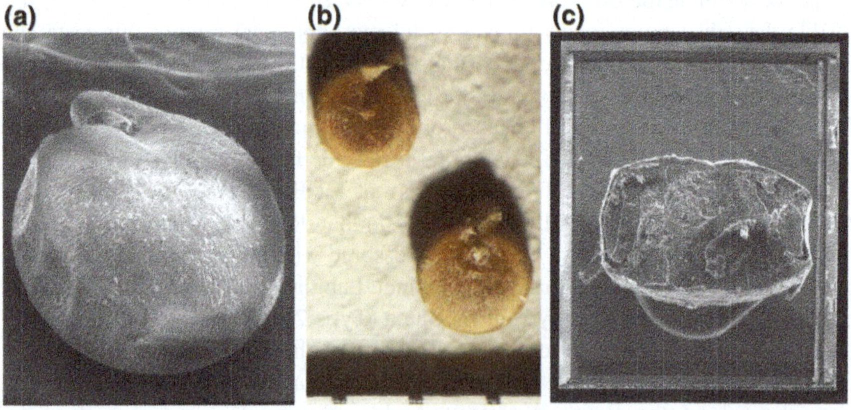

Fig. 3.3 **a** Scanning electron micrograph at 40 × magnification of desiccated 2000-year-old, thin-testa chenopod fruit (seed covered by pericarp) from the Edens Bluff site, Benton County, Arkansas. University Museum, University of Arkansas, accession number 32-3-391. **b** Pale chenopod fruits from Holman Shelter, Madison County, Arkansas, approximately 1000 years old. University Museum, University of Arkansas, accession number 34-22-3. **c** Scanning electron micrograph at 45 × magnification of cross section of pale chenopod fruit from Cow Ford, Benton County, Arkansas. University Museum, University of Arkansas accession number 32-17-22

were present in curated collections from the Riverton site in southern Illinois, excavated by Howard Winters during the 1960s. In 1963, Richard Yarnell collected and floated 50 samples from Riverton, giving us one of the first flotation-recovered assemblages of ancient plant remains from the Americas. Some samples came from contexts including middens directly adjacent to burned clay floors interpreted as "prepared house platforms" (Smith and Yarnell 2009, p. 6562). Hundreds of seeds—including 540 uncharred, "bone-colored" chenopod specimens—were remarkably well preserved due to the clay deposits, allowing documentation of intact pericarp still surrounding the inner epiderm layer and the space where the outer epiderm would have been had it not been reduced to nothingness as a result of domestication.

The presence of pale chenopod in North America more than 1000 years earlier than previously known is terribly exciting, but just as important is the fact that it is grouped at Riverton with lesser numbers of thin-testa *C. berlandieri* ssp. *jonesianum*, thicker-testa chenopod seeds that probably represent a weedy companion, bottle gourd rind (*Lagenaria siceraria*), domestic-sized sunflower (*Helianthus annuus* var. *macrocarpus*) and marshelder seeds (*Iva annua* var. *macrocarpa*), and squash rind (*Cucurbita pepo*). The Riverton site is the earliest North American setting in which so many known and potential crops were grown together as a complex by low-level food producers. Smith and Yarnell (2009) stress the importance of the fact that innovations in food production took place here in resource-rich river valleys subject to no discernible population packing or political pressure, by people also harvesting large quantities of nuts, deer, fish, and other animals.

Just as the methodology originally applied in North America has been utilized in Mexico (Gordon 2006) and South America (Bruno 2006; Langlie et al. 2011), those of us working in the U.S. can now learn from improvements and insights made by our Andeanist colleagues. The early presence of two cultivated chenopod types—pale and black—at Riverton makes it important to look more closely for morphological variability across space and through time in the Eastern Woodlands. As at La Barca, we might expect that early farmers bred cultivars that were either better adapted to local soil and climatic conditions or were preferred for reasons such as color, taste, or cooking properties. Variability has been noted, but we are now in a far better position to apply digital imaging, more standardized measurement, and multiple attribute analysis to key assemblages.

For example, an intriguing assemblage from Cahokia's sub-Mound 51, a feasting deposit dating to A.D. 1050–1100, includes both charred and uncharred material, the latter having been preserved unburned due to rapid, deep burial (Pauketat et al. 2002). When Fritz and students at Washington University at St. Louis examined the seed-rich samples from this deposit, we hoped to find pale chenopod fruits. However, early Cahokia's cultigen chenopod—judging by this deposit—all seemed to have been black and *chia*-like (Fritz 2000; Roberts 1996). Still, the thin-testa specimens are quite large reaching 2.2 mm in diameter. Other attributes including beak prominence and details of testa texture and thickness need to be studied using this and numerous other well-preserved collections. If a distinct variety can be documented for the sub-Mound 51 deposit, it may help us reveal ritual and social dimensions of large-scale gatherings in Cahokia's Grand Plaza.

Ancient and Modern DNA

While there has been no debate that the origin of quinoa was in South America, when *Chenopodium* emerged as a potential crop grown in eastern North America, the debate focused on whether it was independently domesticated or introduced as a cultigen from Mesoamerica (Fritz 1984, 1986; Gilmore 1931). In recent years, the archaeological record itself has been seen as supporting independent domestication. More and more flotation samples from sites in the U.S. Midwest Riverine Area yielded chenopod seeds, projecting a history of: (a) early harvesting of morphologically wild populations, followed by selection of seeds with high proportions of thin seed coats; (b) the presence of seeds with intermediate testa thicknesses representing a probable weedy companion that evolved in the agroecological niche; and (c) the existence of both pale and black, thin-testa morphotypes. This is very different from the situation in Mexico, where evidence for pre-Colonial cultigen chenopod is elusive, in spite of Mesoamerica's rich history as a center of agricultural origins (McClung de Tapia and Rios-Fuentes 2006).

Recent analysis of ancient and modern DNA appears to confirm the independent domestication of *Chenopodium* in eastern North America. Kistler and Shapiro (2011) successfully isolated chloroplast plastid DNA from uncharred archaeological specimens from Cloudsplitter and Haystack rockshelters in eastern Kentucky and from Holman Shelter in Northwest Arkansas. Kistler and Shapiro compared the ancient haplotypes to those of modern wild *C. berlandieri* from eastern North America (17 accessions, including some collected outside the range of EAC agriculture) and from *C. berlandieri* ssp. *nuttalliae* from Mexico (6 accessions). In addition, five samples of modern *C. berlandieri* ssp. *zshackei* and four of *C. berlandieri* ssp. *sinuatum* from western North America were analyzed, and one *C. album* was included as an outgroup. Out of 44 single ancient seed samples, sufficient DNA for amplification and sequencing came from 12 seeds, some pale and some dark thin-testa. All of these conformed genetically to the wild eastern North American haplotype pattern, which was distinct from that of all modern Mexican cultigens. Kistler and Shapiro (2011, p. 3552) conclude that, "chenopod was locally domesticated in eastern North America from native wild populations independent of the cultivated Mexican lineage." This provides "compelling support for the development of an entirely indigenous agricultural complex in ENA."

Independent phylogenetic analysis of DNA—both nuclear and plastid—from modern cultigen chenopod populations and their wild relatives is being done at the University of Wisconsin by Brian Walsh and Eve Emshwiller. Preliminary results reported in 2011 are consistent with independent domestication in Mesoamerica and South America, and future ancient DNA work is planned to further refine the phylogenetic origins of archaeological domestic forms in eastern North America (Walsh and Emshwiller 2011).

While there is little doubt that quinoa was domesticated in South America, its wild progenitor and the specific region (or possibly regions) where domestication took place are still undetermined. Because of its growing prominence as a food crop, the majority of genetic research in quinoa has been motivated by possibilities for cultivar improvement and adaptability. Many of these studies, reviewed thoroughly by Fuentes and Zurita-Silva (2013), have focused on the evolution and genetic basis of agronomically important performance traits (Balzotti et al. 2008; Maughan et al. 2009; Reynolds 2009), population structure and diversity with emphasis on cultivated forms (Christensen et al. 2007; Costa Tártara et al. 2012; delCastillo et al. 2007; Fuentes et al. 2009; Wilson 1988, 1990), and characterization of genome structure and arrangement (Bhargava et al. 2006, 2007; Jarvis et al. 2008; Maughan et al. 2004, 2006; Palomino et al. 2008), largely with respect to breeding goals. Recently, quinoa researchers have integrated genomic approaches such as whole transcriptome and high-throughput expressed sequence tag (EST) sequencing, and high-throughput genotyping (Maughan et al. 2012; Raney et al. 2014; Reynolds 2009). These tools are being used, for example, to analyze gene expression under varying field conditions and among lineages differing in growth attributes, as well as to improve models of genome structure and inheritance, the latter being especially complex issues in tetraploid quinoa. These studies signal a shift toward characterizing quinoa's adaptive pathways and population structure at the whole genome scale, and the resulting datasets will likely also prove useful in refining our understanding of the emergence of domestic forms in South America.

Comparing and Contrasting Sociocultural Contexts

Because of these advances, we are currently situated to better understand the sociopolitical contexts of chenopod domestication and its cultivation history in both ENA and the Andean *altiplano*[1]. While chenopods share a common morphology and ecology, there are interesting contrasts in the roles that they played as crops and foods in cultural and political developments of the two regions.

Eastern North America

Late Archaic (c. 4000–1000 B.C.) North American societies in the Midwest riverine area established settlements along major and minor river valleys as well as in

[1]The Andean region is expansive, and domesticated chenopods likely had distinct trajectories in its subregions. We focus here on our primary research areas in the Bolivian *altiplano* (a high plain that runs between the eastern and western Andes ranges).

upland settings. Tracts of oak-hickory forests, savannahs, and bottomland terraces were managed by fire and other strategies to enhance productivity and hunting efficiency (Delcourt and Delcourt 2004; Smith 2011). Exchange networks facilitated the spread of exotic objects, raw materials, ideas, seeds, and people themselves (Jefferies 1996). As summarized by Smith (2011, p. S481), "... between 5000 and 3500 BP the oak-savannah and oak-hickory forest regions were inhabited by a large number of small autonomous societies, some if not all of which were experimenting to various degrees with the cultivation of local seed plants and sharing their success and failure, as well as their seed stores, along well-established networks of interaction." Domesticated bottle gourds were present across the region by this time, and the native eastern *Cucurbita pepo* ssp. *ovifera* had been domesticated and distributed widely, as had sunflowers and the closely related crop known as marshelder (Rieseberg and Harter 2006; Smith 2006a, 2014). Chenopod entered plots of open, disturbed, enriched soil in or near settlements where early low-level food producers closely observed the properties of a growing number of seed-bearing plants and selected those most attractive to them for storage, propagation, and geographic spread.

Cultural contexts of initial chenopod domestication appear relatively nonhierarchical and disconnected from ritual activities, as far as we can discern. However, by 500 B.C., during the Early Woodland period, chenopod and other members of the EAC had become increasingly visible at places such as Salts and Mammoth Caves, Kentucky, where cavers seeking spiritually charged minerals and underground experiences left direct dietary evidence in the form of paleofeces demonstrating that chenopod contributed heavily to their diets (Crothers 2012; Gardner 1987; Yarnell 1974). Middle Woodland peoples (c. 300 B.C.–A.D. 400) who participated in the construction of elaborate Hopewellian mortuary earthworks and exchanged beautifully crafted objects made of exotic stone, shell, and copper had further increased reliance on cultigen chenopod and other EAC crops (Fritz 1993; Fritz and Smith 1988; Smith 1992a, 2006a, b). Masses of charred chenopod seeds have been recovered from pits dating to the Late Woodland period (c. A.D. 400–1000), a time of less obvious pan-regional ritual display, but significant nonetheless for demographic growth preceding the rise of Mississippian mound centers (Johannessen 1993; Simon 2000; Simon and Parker 2006).

Mississippians (c. A.D.1000–1550) are generally categorized as maize-based farmers under whose influence most of the native EAC crops declined in importance, but the archaeobotanical record from the Central Mississippi Valley—especially the American Bottom area where Cahokia Mounds is located—shows unambiguous evidence for intensification of chenopod, maygrass (*Phalaris caroliniana*), erect knotweed (*Polygonum erectum*), and little barley (*Hordeum pusillum*) along with maize at the end of the first millennium A.D. (Lopinot 1997; Simon and Parker 2006). The contents of the sub-Mound 51 feasting pit at the edge of Cahokia's Grand Plaza, discussed above, attest to the continuing popularity of chenopod and other EAC crops in communal gatherings during the climax of this extremely complex civic-ceremonial center (Pauketat et al. 2002). Surprisingly, maize was poorly represented in this enormous deposit. Not until the greater

Cahokia region was mostly depopulated 600–700 years ago did the native seed crops drop out of the agricultural system that had for centuries sustained the largest, densest, and most politically complex center north of Mesoamerica.

Causes for the decline of chenopod production in eastern North America after c. A.D. 1200 continue to be debated and may never be fully understood. A hemispherical view that encompasses the trajectory of quinoa through the present day raises a red flag against falling back on the suspiciously ethnocentric explanation that *C. berlandieri* ssp. *jonesianum* and other small native grains were inherently inferior to maize in productivity, taste, storability, or other qualities that must have mattered to Mississippian people, including chiefs or priests, during the first half of the second millennium BP. Sociopolitical factors cannot be discounted, however, and should be considered along with ecological ones. Maize did rise to economic dominance; two of the EAC crops—sunflowers and eastern ovifera squashes—were still widely grown and eaten; and, after A.D. 1200, common beans (*Phaseolus vulgaris*) were finally incorporated into late prehistoric farming systems in the Eastern Woodlands. In contrast, chenopod, maygrass, marshelder, erect knotweed, and little barley disappeared as crops without written documentation, although an early eighteenth century French description of a plant called *choupichoul*, sown casually by Natchez Indians in mudflats, is likely to be a reference to chenopod (Smith 1992b).

As part of the shift to a system that focused on maize, beans, and squashes, farmers increasingly concentrated on individual plants that grew in discrete clusters, a classic pattern being the "hills" of intercropped Three Sisters agriculture as practiced by historic Iroquois-speaking tribes (Mt. Pleasant 2006). Small grains that were probably broadcast by hand rather than planted in holes dug by digging sticks or hoes seem to have become relegated to peripheral patches in regions where they had once played a major economic role. Furthermore, dramatic population shifts occurred across the Midwest prior to A.D. 1492, and the impacts of pre-Colonial emigration and relocation must have disrupted both agricultural traditions and cuisines even before Old World invaders and diseases began wreaking havoc. The American Bottom, former home to thousands of participants in the Cahokian social system, was largely depopulated after A.D. 1350, as were adjoining expanses of what archaeologists call the Vacant Quarter (Edging 2007). Whatever the causes of chenopod's decline, this issue warrants further examination by researchers studying the complex dynamics of past foodways and traditional farming systems.

The Andean Altiplano

Hunters and foragers spread out across the arid highland Andes during the late Archaic period (3200–1800 B.C.) as the environment became more suitable for human habitation with warmer temperatures, increased rainfall, and patches of vegetation flourishing near lacustrine and riverine zones (Rigsby et al. 2003). These early inhabitants seasonally occupied logistical encampments across diverse ecological areas (Aldenderfer 1989; Capriles Flores 2014). Intensification of hunting wild camelids led to management of these herds, and llamas and alpacas were fully domesticated and incorporated into the subsistence economy of transient groups around 2200 B.C. (Mengoni Goñalons and Yacobaccio 2006, p. 239). To water and feed their animals, camelid herders moved between pockets of ecologically diverse areas along the shores of rivers and lakes. Due to its weedy tendencies and role as a food for camelids, chenopods likely proliferated in the fertilized soils of corrals and in disturbed human encampments. It is within these anthropogenic areas that chenopods were likely brought under cultivation as the result of a mutualistic relationship between camelids, chenopods, and humans (Kuznar 1993; Pearsall 1992). Along with chenopods, several local tuber species were also domesticated including another crop of modern worldwide importance, the potato (*Solanum tuberosum* L.) (Hastorf 2006; Spooner et al. 2005).

Chenopods are estimated to have been domesticated toward the end of the Archaic period as early as 3000 B.C. (Bruno 2006, p. 43), but currently the only direct evidence of the domesticated form dates to the Early Formative period (around 1500 B.C.) at several sites in the Lake Titicaca basin (Bruno 2001; Eisentraut 1998). It was during the Formative period from 1500 B.C. to A.D. 400 that many of the social consequences of plant and animal domestication appear such as: the transition from transhumance to sedentary village life (Bandy 2004), increased craft specialization with common iconographic themes (Browman 1980), early signs of status differentiation (Rose 2001), and the development of unified religious traditions (Burger et al. 2000). Throughout the *altiplano* as the Formative period progressed, economically specialized farmers, fishers, and pastoralists all complemented their diets with domesticated chenopods (Bruno 2006; Bruno and Whitehead 2003; Eisentraut 1998; Langlie 2011; Langlie et al. 2011). Chenopods appear regularly in the archaeological record both in household food middens but also in unique ritual and political contexts. High densities of a wild chenopod, possibly *C. pallidicaule* (relative of the domesticate kañawa/cañihua), are found burned in situ above clay floors in Middle Formative sunken courts at the site of Kala Uyuni (the Achachi Coa Collu sector), on the southern shores of Taraco Peninsula, Bolivia (Bruno 2008, pp. 308, 309). At the site of Chiripa, on the northern shore of the Taraco Peninsula, residents stored large quantities of quinoa in bins of small structures that were part of a platform mound (Bruno and Whitehead 2003; Towle 1961). The Chiripa mound is one of the most prominent examples of Formative period corporate architecture in the region (Bandy 2001) and was continuously used for political and ceremonial purposes for generations (Hastorf 2003). The storage of quinoa in this location indicates that it was more than a mundane foodstuff during Formative period times.

The complex economic and political interactions that began in the Formative period coalesced into the first state in the southern Andes, Tiwanaku (A.D. 500–1000) (Janusek 2008; Kolata 1993). The civic-ceremonial center of the state was located just 20 km southeast of the shores of Lake Titicaca and the Taraco Peninsula, but the state's influences reached southern Peru, northern Chile and Argentina, and central Bolivia. The economic foundations of the state involved both extensive trade networks aided by llama caravans (Browman 1980), as well as intensified agriculture (Kolata 1986). Tiwanaku leaders coordinated the expansion of raised field agriculture along the shores of Lake Titicaca (Janusek and Kolata 2004; Kolata 1986, 1991). These productive agricultural systems ensured high yields of chenopods and tubers necessary to sustain growing populations. Imported maize took on an important new political role in the region with the Tiwanaku state (Hastorf et al. 2006). Fermented maize beer was an essential part of state ceremonies and celebrations as evidenced by specialized drinking and fermentation vessels (Goldstein 2003), as well as increased C4 levels in human isotopes (Berryman 2010). Archaeobotanical studies show, however, that local chenopods and tubers continued to provide the basis of *altiplano* diets even at the urban center of Tiwanaku (Wright et al. 2003).

Around A.D. 1100, Tiwanaku collapsed and increased social tensions and climatic variability erupted into all out warfare throughout the southern highlands during the Late Intermediate period from A.D. 1000 to 1450 (Arkush 2008). People abandoned the Tiwanaku center, and populations dispersed across the landscape. For defense, warring groups strategically coalesced in hillforts (Albarracín-Jordan 1992, pp. 227–284; Arkush 2011; Bauer and Stanish 2001; Stanish 1994, p. 322). Recent archaeobotanical analysis indicates that these groups relied heavily on chenopods for subsistence (Langlie and Arkush 2016). Specifically, several dense caches of charred chenopods were found during excavations at the Late Intermediate period site Ayawiri, located west of Lake Titicaca near Puno, Peru. These caches were found in various household contexts such as pits below house floors and cooking hearths. These data indicate that even though there was climatic instability and minimal political or religious continuity in the *altiplano* after A.D. 1100, farmers maintained their long-held agricultural traditions and foodways. Additional evidence indicates this was the case throughout the far southern *altiplano* as well. Examination of uncarbonized quinoa stores in the Lípez region of Bolivia (near the Argentine border) shows that LIP farmers were cultivating a wide number of varieties for distinct culinary purposes (López and Nielsen 2012).

Around A.D. 1450, the Inca conquered the Lake Titicaca basin. Socially, politically, and economically the entire Andean region was integrated into the Inca Empire until Spanish forces took over approximately 90 years later (Rowe 1945). When the Spanish arrived, they documented the important role of quinoa to the Inca economy. For example, Betanzos (1996 [1557]) noted that when Topa Inca Yupanque consolidated the Inca Empire in the latter half of the fifteenth century, he ordered all the lords who oversaw the hinterlands to construct storehouses (known as *qollqa*) in the capital city Cuzco, and to fill these granaries with dried provisions brought in from those regions, particularly crops. Quinoa was mentioned as one of these staple finance foods, supplying sustenance for common city dwellers, elite royal politicians, warriors, and all who lived in the capital city. The Inca also demanded that the conquered regions provide laborers to help produce various crops on provincial fields, including quinoa (D'Altroy and Hastorf 1992, pp. 264–273). Several archaeologists have investigated the large number of storehouses found in Cuzco as well in the provincial centers (Bauer 2004, pp. 96–97; D'Altroy and

Hastorf 1984; LeVine 1992; Morris 1976), and thousands of charred quinoa grains have been archaeologically recovered from excavated storehouses (D'Altroy and Hastorf 1984). The tribute collection and redistribution of quinoa and other foods by the Inca provided payment, sustenance, and support for state-financed activities throughout the Andes (D'Altroy and Hastorf 1984; Earle 1992, p. 335).

After the Spanish conquest of the Andes, quinoa continued to be a mainstay in the diet for indigenous inhabitants of the region. Perhaps this was due to quinoa's central importance not just as a vigorous crop and nutritious food, but also central to Andean rituals as a fermented beverage. For example, the Jesuit priest Father Bernabe Cobo observed while visiting the Andes in the seventeenth century that quinoa was a supremely important as chicha beer. He elaborates that at the time of conquest chicha (whether made from quinoa, maize, or molle berries) was "the height of their glory… (Andean people) never celebrate an event, whether joyful or sad, in any way other than by dancing and drinking to excess" (Cobo 1979, p. 135). In the nineteenth century, when Prussian geographer and naturalist Alexander von Humboldt traveled through Columbia, he discerned that quinoa was for indigenous people of the Andes what "wine was to the Greeks, wheat to the Romans, cotton to the Arabs" (Popenoe et al. 1989, p. 151). Although the Spanish conquerors found quinoa to be an exotic product, and later considered it an "Indian" foodstuff, its cultivation was never prohibited, and quinoa continued to provide nutritive sustenance for families (Hunziker 1952). In modern times, chenopods continue to be a pillar of the Andean *altiplano* diet, with quinoa only recently gaining worldwide popularity.

On the heels of the European conquest of the Americas, New World domesticates such as maize, potatoes, chili peppers, and tomatoes were brought to Europe and integrated into Old World cuisines. However, quinoa did not become an important part of this Columbian Exchange. Apparently the Spanish tried to grow quinoa seeds in Spain, but they failed because the seeds "arrived dead" (Tapia 2015, p. 4). Furthermore, Europeans such as Cobo (1945 [1663]) confused quinoa with the native weedy amaranth that grew on the Iberian Peninsula. This confusion likely contributed to the worldwide obscurity of quinoa outside of South America throughout the colonial era (Tapia 2015, p. 4).

Conclusions/Current and Future Directions

Chenopodium is now an icon for revival of "lost" crops (Gremillion 2014; National Research Council 1989), enabling us to learn about past foodways and to eat more healthy food today. Although this food was not "lost" to the Andeans who still eat it and now share it with the world, archaeobotanists continue to provide key insights into its domestication and prehistory in the Andes, and perhaps more significantly in eastern North America, where this crop was genuinely lost. Morphological studies integrating new techniques with old-fashioned microscopy of exceptionally preserved museum collections as well as newly excavated ones not only are refining the story of when and where chenopods were domesticated, but are also revealing the amazing diversity of varieties and subspecies cultivated by farmers north and south. Future research into how this diversity played out in different culinary contexts promises to enrich our understanding of how these crops were shaped by and contributed to the transformations of social and political life of these ancient societies. The molecular secrets of both ancient and modern chenopods are being unraveled, and their anatomical variability categorized.

The comparison of chenopods in eastern North America and the Andean *altiplano* reveals important differences between the trajectories of this food in both regions; there are, however, some interesting similarities. In both areas, chenopods likely introduced themselves as camp followers to foraging people's settlements in early gardens and, in the Andes, in corrals. As farmers encouraged the plant in garden plots and courtyard processing areas, it appears to have become an important contributor to both daily and special meals of early complex societies in both regions. The introduction of maize into the Andes, particularly as a food of ceremony and political clout, presented a challenge to the role of quinoa; yet, quinoa appears to have remained important to local farmers and managed to make its way into some political feasts. It is likely that quinoa's status as a reliable, decentralized foodstuff that could be cultivated at higher elevations than possible for maize allowed it to thrive during the Late Intermediate Period of the Andes. In eastern North America, the domesticated *Chenopodium* species did not enjoy a similar, strategic advantage; several centuries after maize was intensified, but before the European intrusions, the crop had lost its place as one of the most important grains produced across the Midwest and northern Southeast.

Altiplano farmers in the Andes also faced imperial incursions into their lives, first with the Inca and then the Spanish. The quinoa crop lost some of its acreage to barley and other Eurasian crops, but it continued to play a role in indigenous households. Thanks to these farmers, we can enjoy it today. Although it may seem unlikely, and would probably be quite challenging, the re-domestication of eastern North American *Chenopodium berlandieri* is not beyond the capabilities of modern researchers. The ecological and economic implications of this revival in our modern sociopolitical contexts would certainly add another interesting chapter to the long history of domesticated chenopods.

Acknowledgements We are deeply indebted to Christine Hastorf for her contributions that not only generated a great deal of the archaeological material we discuss, but also established the framework for integrating archaeobotanical data with everything else. Chenopod research benefits from Christine's vision of food and regional cuisines as central to and inseparable from considerations of identity, status, exchange, and the development of politically complex societies. As a project P.I., lab director, teacher, and theoretical innovator, Christine influenced, inspired, and enabled much of this research. We also sincerely thank Dr. Lynne Rouse for drafting the Fig. 3.1 map.

References

Albarracín-Jordan, J. (1992). *Prehispanic and early colonial settlement patterns in the lower Tiwanaku valley, Bolivia*. Ph.D. Dissertation. Dallas: Department of Anthropology, Southern Methodist University.

Aldenderfer, M. S. (1989). Archaic period in the South Central Andes. *Journal of World Prehistory, 3*(2), 117–158.

Arkush, E. N. (2008). War, causality, and chronology in the Titicaca Basin. *Latin American Antiquity, 19*(4), 339–373.

Arkush, E. N. (2011). *Hillforts of the ancient Andes: Colla warfare, society, and landscape*. Gainesville: University Press of Florida.

Balzotti, M. R. B., Thornton, J. N., Maughan, P. J., McClellan, D. A., Stevens, M. R., Jellen, E. N., et al. (2008). Expression and evolutionary relationships of the Chenopodium quinoa 11S seed storage protein gene. *International Journal of Plant Sciences, 169*, 281–291.

Bandy, M. S. (2001). *Population and history in the ancient Titicaca Basin*. Ph. D. Dissertation. Berkeley: Department of Anthropology, University of California.

Bandy, M. S. (2004). Fissioning, scalar stress, and social evolution in early village societies. *American Anthropologist, 106*(2), 322–333.

Bauer, B. S. (2004). *Ancient Cuzco: heartland of the Inca*. Austin: University of Texas Press.

Bauer, B. S., & Stanish, C. (2001). *Ritual and pilgrimage in the ancient Andes: the Islands of the Sun and the Moon*. Austin: University of Texas Press.

Berryman, C. A. (2010). *Food, feasts, and the construction of identity and power in ancient Tiwanaku: a bioarchaeological perspective*. Ph. D. Dissertation. Nashville, TN: Department of Anthropology, Vanderbilt University.

Betanzos, J. (1996). *Narrative of the Incas: from the Palma de Mallorca Manuscript*, 1557 (trans: Hamilton, R., & Buchanan, D.). Austin: University of Texas Press.

Bhargava, A., Shukla, S., & Ohri, D. (2006). Karyotypic studies on some cultivated and wild species of Chenopodium (Chenopodiaceae). *Genetic Resources and Crop Evolution, 53*, 1309–1320.

Bhargava, A., Shukla, S., & Ohri, D. (2007). Genome size variation in some cultivated and wild species of Chenopodium (Chenopodiaceae). *Caryologia, 60*, 245–250.

Browman, D. L. (1980). Tiwanaku expansion and altiplano economic patterns. *Estudios Arqueologicos, 5*, 107–120.

Bruno, M. C. (2001). *Formative Agriculture? The status of Chenopodium domestication and intensification at Chiripa, Bolivia* (1500 *B.C.–A.D.*100). Master's Thesis. St. Louis, MO: Department of Anthropology, Washington University.

Bruno, M. C. (2006). A morphological approach to documenting the domestication of Chenopodium in the Andes. In M. A. Zeder, D. G. Bradley, E. Emshwiller, & B. D. Smith (Eds.), *Documenting domestication: new genetic and archaeological paradigms* (pp. 32–45). Berkeley: University of California Press.

Bruno, M. C. (2008). *Waranq Waranqa: Ethnobotanial Perspectives on agricultural intensification in the Lake Titicaca Basin (Taraco Peninsula, Bolivia)*. Ph.D. Dissertation. St. Louis: Department of Anthropology, Washington University.

Bruno, M. C., Rojas, W., & Pinto, M. (2013). *Morfología de semillas de cañahua (Chenopodium pallidicaule Aellen) silvestres y domesticadas: hacia un mejor entendimiento de los procesos de domesticación*. Ibarra, Ecuador: Paper presented at the IV Congreso Mundial de Quinua.

Bruno, M. C., & Whitehead, W. T. (2003). Chenopodium cultivation and formative period agriculture at Chiripa, Bolivia. *Latin American Antiquity, 14*, 339–355.

Burger, R. L., Mohr-Chávez, K. L., & Chávez, S. J. (2000). Through the glass darkly: prehispanic obsidian procurement and exchange in southern Peru and northern Bolivia. *Journal of World Prehistory, 14*, 267–362.

Capriles Flores, J. M. (2014). *The economic organization of early camelid pastoralism in the Andean highlands of Bolivia*. Oxford: British Archaeological Reports.

Christensen, S. A., Pratt, D. B., Pratt, C., Nelson, P. T., Stevens, M. R., Jellen, E. N., et al. (2007). Assessment of genetic diversity in the USDA and CIP-FAO international nursery collections of quinoa (Chenopodium quinoa Willd.) using microsatellite markers. *Plant Genetic Resources: Characterization and Utilization, 5,* 82–95.

Cobo, B. (1945). *Historia del nuevo mundo, Biblioteca de autores españoles*. Madrid: Editorial Atlas.

Cobo, B. (1979). *History of the Inca Empire* (trans: Hamilton, R.). Austin: University of Texas Press.

Costa Tártara, S. M., Manifesto, M. M., Bramardi, S. J., & Bertero, H. D. (2012). Genetic structure in cultivated quinoa (Chenopodium quinoa Willd.), a reflection of landscape structure in Northwest Argentina. *Conservation Genetics, 13,* 1027–1038.

Crothers, G. M. (2012). Early Woodland ritual use of caves in eastern North America. *American Antiquity, 77,* 524–541.

D'Altroy, T. N., & Hastorf, C. A. (1984). The distribution and contents of Inca state storehouses in the Xauxa region of Peru. *American Antiquity, 49,* 334–349.

D'Altroy, T. N., & Hastorf, C. A. (1992). The architecture and contents of Inka State storehouses in the Xauxa Region of Peru. In T. Y. LeVine (Ed.), *Inka storage systems* (pp. 259–286). Norman: University of Oklahoma Press.

del Castillo, C., Winkel, T., Mahy, G., & Bizoux, J. (2007). Genetic structure of quinoa (Chenopodium quinoa Willd.) from the Bolivian altiplano as revealed by RAPD markers. *Genetic Resources and Crop Evolution, 54,* 897–905.

Delcourt, P. A., & Delcourt, H. R. (2004). *Prehistoric native Americans and ecological change: human ecosystems in eastern North America since the Pleistocene*. Cambridge: Cambridge University Press.

DeWet, J. M. J., & Harlan, J. (1975). Weeds and domesticates: evolution in the man-made habitat. *Economic Botany, 29,* 99–107.

Earle, T. K. (1992). Storage and the Inka imperial economy: archaeological research. In T. Y. LeVine (Ed.), *Inka Storage Systems* (pp. 327–342). Norman: University of Oklahoma Press.

Edging, R. (2007). The vacant quarter hypothesis: a survivor's story. *Missouri Archaeologist, 68,* 59–93.

Eisentraut, P. J. (1998). *Macrobotanical remains from southern Peru: a comparison of Late Archaic-early formative period sites from the Puna and Suni zones of the Western Titicaca Basin*. Ph. D. Dissertation. Santa Barbara, CA: Department of Anthropology, University of California.

Fritz, G. J. (1984). Identification of cultigen amaranth and chenopod from rockshelter sites in Northwest Arkansas. *American Antiquity, 49,* 558–572.

Fritz, G. J. (1986). *Prehistoric Ozark agriculture: the University of Arkansas rockshelter collections*. Ph. D. Dissertation. Chapel Hill, NC: Department of Anthropology, University of North Carolina at Chapel Hill.

Fritz, G. J. (1993). Early and Middle Woodland period paleoethnobotany. In C. M. Scarry (Ed.), *Foraging and farming in the eastern Woodlands* (pp. 39–56). Gainesville: University Press of Florida.

Fritz, G. J. (2000). Food and ceremonial plants from Sub-Mound 51 at Cahokia. Report submitted to Timothy Pauketat, P.I. of NSF-Sponsored Early Cahokia Project. Ms. on file at Washington University in St. Louis, Paleoethnobotanical Laboratory.

Fritz, G. J., & Smith, B. D. (1988). Old collections and new technology: documenting the domestication of Chenopodium in eastern North America. *Midcontinental Journal of Archaeology, 13,* 3–27.

Fuentes, F. F., Martinez, E. A., Hinrichsen, P. V., Jellen, E. N., & Maughn, P. J. (2009). Assessment of genetic diversity patterns in Chilean quinoa (Chenopodium quinoa Willd.) germplasm using multiplex fluorescent microsatellite markers. *Conservation Genetics, 10,* 369–377.

Fuentes, F. F., & Zurita-Silva, A. (2013). Molecular studies. In A. Bhargava & S. Srivastava (Eds.), *Quinoa: Botany, production and uses* (pp. 168–184). Boston: CABI.

Gardner, P. S. (1987). New evidence concerning the chronology and paleoethnobotany of Salts Cave, Kentucky. *American Antiquity, 52,* 358–366.
Gilmore, M. R. (1931). Vegetal remains of the Ozark Bluff-Dweller culture. *Papers of the Michigan Academy of Science, Arts, and Letters, 14,* 38–102.
Goldstein, P. S. (2003). From stew-eaters to maize-drinkers, the chicha economy and the Tiwanaku expansion. In T. L. Bray (Ed.), *The archaeology and politics of food and feasting in early states and empires* (pp. 143–172). New York: Springer.
Gordon, A. G. (2006). *Domesticated Chenopodium in North America: comparing the past and the present.* Ph. D. Dissertation. St. Louis, MO: Department of Anthropology, Washington University.
Gremillion, K. J. (1993). The evolution of seed morphology in domesticated Chenopodium: an archaeological case study. *Journal of Ethnobiology, 13,* 149–169.
Gremillion, K. J. (2014). Goosefoot (Chenopodium). In P. E. Minnis (Ed.), *New lives for ancient and extinct crops* (pp. 44–64). Tucson: University of Arizona Press.
Hastorf, C. A. (2003). Community with the ancestors: ceremonies and social memory in the middle formative at Chiripa, Bolivia. *Journal of Anthropological Archaeology, 22,* 305–332.
Hastorf, C. A. (2006). Domesticated food and society in early coastal Peru. In W. Balée & C. L. Erickson (Eds.), *Time and complexity in historical ccology: studies in the Neotropical lowlands* (pp. 87–126). New York: Columbia University Press.
Hastorf, C. A., Whitehead, W. T., Bruno, M. C., & Wright, M. (2006). The movements of maize into middle horizon Tiwanakau, Bolivia. In R. Tykot, J. Staller, & B. Benz (Eds.), *Histories of Maize: multidisciplinary approaches to the prehistory, linguistics, biogeography, domestication, and evolution of Maize* (pp. 429–448). Oxford: Academic Press.
Hunziker, A. (1952). *Los pseudocereales de la agricultura indígena de América.* Buenos Aires: Acme Agency.
Janusek, J. W. (2008). *Ancient Tiwanaku.* Cambridge: Cambridge University Press.
Janusek, J. W., & Kolata, A. L. (2004). Top-down or bottom-up: rural settlement and raised field agriculture in the Lake Titicaca Basin, Bolivia. *Journal of Anthropological Archaeology, 23,* 404–430.
Jarvis, D. E., Kopp, O. R., Jellen, E. N., Mallory, M. A., Pattee, J., Bonifacio, A., et al. (2008). Simple sequence repeat marker development and genetic mapping in quinoa (Chenopodium quinoa Willd.). *Journal of Genetics, 87,* 39–51.
Jefferies, R. W. (1996). The emergence of long distance trade networks in the Southeastern United States. In K. Sassaman & D. Anderson (Eds.), *Archaeology of the mid-Holocene Southeast* (pp. 222–234). Gainesville: University Press of Florida.
Johannessen, S. (1993). Farmers of the Late Woodland. In C. M. Scarry (Ed.), *Foraging and farming in the eastern Woodlands* (pp. 57–77). Gainesville: University Press of Florida.
Kistler, L., & Shapiro, B. (2011). Ancient DNA confirms a local origin of domesticated chenopod in eastern North America. *Journal of Archaeological Science, 38,* 3549–3554.
Kolata, A. L. (1986). The agricultural foundations of the Tiwanaku State: a view from the Heartland. *American Antiquity, 51*(4), 748–762.
Kolata, A. L. (1991). The technology and organization of agricultural production in the Tiwanaku state. *Latin American Antiquity,* 2(2), 99–125.
Kolata, A. L. (1993). *The Tiwanaku: portait of an Andean civilization.* Oxford: Blackwell Press.
Kuznar, L. A. (1993). Mutualism between Chenopodium, herd animals, and herders in the south central Andes. *Mountain Research and Development, 13,* 257–265.
Langlie, B. S. (2011). *A paleoethnobotanical analysis of three formative period Wankarani sites located in the department of Oruro, Bolivia.* M.A. Thesis. St. Louis, MO: Department of Anthropology, Washington University.
Langlie, B. S., & Arkush, E. N. (2016). Managing mayhem: conflict, environment, and subsistence in the Andean Late Intermediate Period, Puno, Peru. In A. VanDerwarker & G. Wilson (Eds.), *The archaeology of food and warfare: food insecurity in prehistory* (pp. 259–290). New York: Springer.

Langlie, B. S., Hastorf, C. A., Bruno, M. C., Bermann, M., Bonzani, R. M., & Castellón Condarco, W. (2011). Diversity in Andean Chenopodium domestication: describing a new morphological type from La Barca, Bolivia 1300-1250 BC. *Journal of Ethnobiology, 31,* 72–88.

LeVine, T. Y. (1992). *Inka storage systems*. Norman: University of Oklahoma Press.

López, M. L., Bruno, M. C., & Planella, M. T. (2015). El género Chenopodium: metodología aplicada a la identificación taxonómica en ejemplares arqueológicos. Presentación de Casos de Estudio de La Región Sur-Andina. In C. Belmar, & V. Lema (Eds.), *Avances y desafíos metodológicos en arqueobotánica: miradas consensuadas y diálogos compartidos desde Sudamérica* (pp. 89–192, Monografías arqueológicas, Facultada Patrimonio, Cultura, y Educación). Chile: Universidad SEK.

López, M. L., & Nielsen, A. E. (2012). Macrorrestos de Chenopodium quinoa Willd. en la plaza de Laqaya (Nor Lípez, Potosí, Bolivia). *Revista Intersecciones en Antropología, 14,* 295–300.

Lopinot, N. H. (1997). Cahokian food production reconsidered. In T. R. Pauketat & T. E. Emerson (Eds.), *Cahokia: domination and ideology in the Mississippian world* (pp. 52–68). Lincoln: University of Nebraska Press.

Maughan, P. J., Bonifacio, A., Jellen, E. N., Stevens, M., Coleman, C. E., Ricks, M., et al. (2004). A genetic linkage map of quinoa (Chenopodium quinoa) based on AFLP, RAPD, and SSR markers. *Theoretical and Applied Genetics, 109,* 1188–1195.

Maughan, P. J., Kolano, B. A., Maluszynska, J., Coles, N. D., Bonifacio, A., Rojas, J., et al. (2006). Molecular and cytological characterization of ribosomal RNA genes in Chenopodium quinoa and Chenopodium berlandieri. *Genome, 49,* 825–839.

Maughan, P. J., Smith, S. M., Rojas-Beltrán, J. A., Elzinga, D., Raney, J. A., Jellen, E. N., et al. (2012). Single nucleotide polymorphisms identification, characterization and linkage mapping in Chenopodium quinoa. *The Plant Genome, 5,* 1–7.

Maughan, P. J., Turner, T. B., Coleman, C. E., Elzinga, D. B., Jellen, E. N., Morales, J. A., et al. (2009). Characterization of salt overly sensitive (SOS1) gene homoeologs in quinoa (Chenopodium quinoa Willd). *Genome, 52,* 647–657.

McClung de Tapia, E., & Rios-Fuentes, J. (2006). *Chenopodium spp.: on the road to domestication in the pre-hispanic Basin of Mexico*. Paper presented at the 71st Annual Meeting of the Society for American Archaeology, Puerto Rico, 26 April.

Mengoni Goñalons, G. L., & Yacobaccio, H. D. (2006). The domestication of South American camelids: a view from the South-Central Andes. In M. A. Zeder, D. G. Bradley, E. Emshwiller, & B. D. Smith (Eds.), *Documenting domestication: new genetic and archaeological paradigms* (pp. 228–244). Berkeley: University of California Press.

Morris, C. E. (1976). The archaeological study of Andean exchange systems. In C. L. Redman, M. J. Berman, E. V. Curtin, W. T. Langhorne Jr., N. M. Versaggi, & J. C. Wanser (Eds.), *Social archaeology: beyond subsistence and dating* (pp. 315–337). New York: Academic Press.

Mt. Pleasant, J. (2006). The science behind the Three Sisters mound systems: an agronomic assessment of an indigeous agricultural system in the Northeast. In J. E. Staller, R. H. Tykot, & B. F. Benz (Eds.), *Histories of Maize: multidisciplinary approaches to the prehistory, linguistics, biogeography, domestication, and evolution of Maize* (pp. 529–538). New York: Academic Press.

National Research Council. (1989). *Lost crops of the Incas: little-known plants of the Andes with promise for worldwide cultivation (Report of an ad hoc panel of the Advisory Committee on Technology Innovation Board on Science and Technology for International Development)*. Washington, D.C.: National Academy Press.

Nordstrom, C. (1990). *Evidence for the domestication of Chenopodium in the Andes (Report to the National Science Foundation, Paleoethnobotany Laboratory Reports #19)*. Berkeley: University of California.

Palomino, G., Hernández, L. T., & Torres, E. (2008). Nuclear genome size and chromosome analysis in Chenopodium quinoa and C. berlandieri subsp. nuttalliae. *Euphytica, 164,* 221–230.

Pauketat, T. R., Kelly, L. S., Fritz, G. J., Lopinot, N. H. E. S., & Hargrave, E. (2002). The residues of feasting and public ritual at early Cahokia. *American Antiquity, 67,* 257–279.

Pearsall, D. (1992). The origins of plant cultivation in South America. In C. W. Cowan & P. J. Watson (Eds.), *The origins of agriculture* (pp. 173–205). Washington, D.C.: Smithsonian Institution Press.
Planella, M. T., Lopez, M. L., & Bruno, M. C. (2014). La Domesticación y Distribución Prehistórica. In D. Bazile, D. Bertero, & C. Nieto (Eds.), *State of the art report of Quinoa in the world in 2013: FAO (Santiago de Chile) y CIRAD*. Montpellier: France.
Planella, M. T., McRostie, V., & Falabella, F. (2010). El aporte arqueobotánico al conocimiento de los recursos vegetales en la población alfarera temprana del sitio El Mercurio. Actas del XVII Congreso Nacional de Arqueología Chilena, pp. 1255–1265.
Planella, M. T., Scherson, R., & McRostie, V. (2011). Sitio El Plomo y nuevos registros de cultígenos iniciales en cazadores del Arcaico IV en Alto Maipo, Chile Central. *Chungará, 43* (2), 189–202.
Popenoe, H., King, S. R., León, J. R., & Kalinowski, L. S. (1989). *Lost crops of the Incas: Little-known plants of the Andes with promise for worldwide cultivation*. Washington, D.C.: National Academy Press.
Raney, J. A., Reynolds, D. J., Elzinga, D. B., Page, J., Udall, J. A., Jellen, E. N., et al. (2014). Transcriptome analysis of drought induced stress in Chenopodium quinoa. *American Journal of Plant Sciences, 5,* 338–357.
Reynolds, D. J. (2009). *Genetic dissection of triterpenoid saponin production in Chenopodium quinoa using microarray analysis*. M.Sc. Thesis. Provo, UT: Brigham Young University.
Rieseberg, L. H., & Harter, A. V. (2006). Molecular evidence and the evolutionary history of the domesticated sunflowers. In T. Motley (Ed.), *Darwin's harvest: new approaches to the origins, evolution, and conservation of crops: a broad taxonomic and geographic survey* (pp. 31–48). New York: Columbia University Press.
Rigsby, C. A., Baker, P. A., & Aldenderfer, M. S. (2003). Fluvial history of the Rio Ilave valley, Peru, and its relationship to climate and human history. *Palaeogeography, Palaeoclimatology, Palaeoecology, 194,* 165–185.
Roberts, K. M. (1996). *Intimations of early Mississippian political, domestic, and ecological environments at Cahokia: an analysis of plant remains from Sub-Mound 51*. M.A. Thesis. St. Louis, MO: Department of Anthropology, Washington University.
Rose, C. E. (2001). *Household and community organization of a Formative Period, Bolivian settlement*. Ph. D. Dissertation. Pittsburgh, PA: Department of Anthropology, University of Pittsburgh.
Rowe, J. H. (1945). Absolute chronology in the Andean area. *American Antiquity, 10,* 265–284.
Simon, M. L. (2000). Regional variations in plant use strategies in the Midwest during the Late Woodland. In T. E. Emerson, D. L. McElrath, & A. C. Fortier (Eds.), *Late Woodland societies: tradition and transformation across the midcontinent* (pp. 37–75). Lincoln: University of Nebraska Press.
Simon, M. L., & Parker, K. E. (2006). Prehistoric plant use in the American Bottom: new thoughts and interpretations. *Southeastern Archaeology, 25,* 212–257.
Smith, B. D. (1985a). Chenopodium berlandieri ssp. jonesianum: evidence for a Hopewellian domesticate from Ash Cave, Ohio. *Southeastern Archaeology, 4,* 107–133.
Smith, B. D. (1985b). The role of Chenopodium as a domesticate in the pre-maize garden systems of the eastern United States. *Southeastern Archaeology, 4,* 51–72.
Smith, B. D. (1992a). Hopewellian farmers of eastern North America. In B. D. Smith (Ed.), *Rivers of change: essays on early agriculture in eastern North America* (pp. 201–248). Washington, D.C.: Smithsonian Institution Press.
Smith, B. D. (1992b). In search of Choupihoul, the mystery grain of the Natchez. In B. D. Smith (Ed.), *Rivers of change: essays on early agriculture in eastern North America* (pp. 239–264). Washington, D.C.: Smithsonian Institution Press.
Smith, B. D. (2006a). Eastern North America as an independent center of plant domestication. *Proceedings of the National Academy of Sciences, 103,* 12223–12228.

Smith, B. D. (2006b). Household, community, and subsistence in Hopewell research. In D. K. Charles & J. E. Buikstra (Eds.), *Recreating Hopewell* (pp. 491–509). Gainesville: University Press of Florida.

Smith, B. D. (2011). The cultural context of plant domestication in eastern North America. *Current Anthropology, 52*(Supplement 4), S471–S484.

Smith, B. D. (2014). The domestication of Helianthus annuus L. (sunflower). *Vegetation History and Archaeobotany, 23,* 57–74.

Smith, B. D., & Yarnell, R. A. (2009). Initial formation of an indigenous crop complex in eastern North America at 3800 B.P. *Proceedings of the National Academy of Sciences, 106,* 6561–6566.

Spooner, D. M., McLean, K., Ramsay, G., Waugh, R., & Bryan, G. J. (2005). A single domestication for potato based on multilocus amplified fragment length polymorphism genotyping. *Proceedings of the National Academy of Sciences, 102,* 14694–14699.

Stanish, C. (1994). The hydraulic hypothesis revisited: Lake Titicaca basin raised fields in theoretical perspective. *Latin American Antiquity, 5,* 312–332.

Tapia, M. E. (2015). The long journey of Quinoa: who wrote its history? In D. Bazile, D. Bertero, & C. Nieto (Eds.), *State of the art report of Quinoa in the world in 2013* (pp. 3–9). Montpellier, France: FAO (Santiago de Chile) y CIRAD.

Towle, M. A. (1961). *The ethnobotany of pre-Columbian Peru (Viking Publications in Anthropology 30).* Chicago: Aldine.

Wahl, H. A. (1952). A preliminary study of the genus Chenopodium in North America. *Bartonia, 27,* 1–46.

Walsh, B., & Emshwiller, E. (2011). *Phylogeny of American Chenopodium species with focus on origins of the domesticated taxa.* St. Louis: Paper presented at the Annual Meeting of the Botanical Society of America.

Wilson, H. D. (1981). Domesticated Chenopodium of the Ozark Bluff Dwellers. *Economic Botany, 35,* 233–239.

Wilson, H. D. (1988). Quinua biosystematics I: domesticated populations. *Economic Botany, 42,* 461–477.

Wilson, H. D. (1990). Quinua and relatives (Chenopodium subsect. Cellulata). *Economic Botany, 44*(suppl), 92–110.

Wilson, H. D., & Heiser, C. B., Jr. (1979). The origin and evolutionary relationships of "Huauzontle" (Chenopodium nuttalliae Safford), domesticated chenopod of Mexico. *American Journal of Botany, 66,* 198–206.

Wright, M. F., Hastorf, C. A., & Lennstrom, H. A. (2003). Pre-Hispanic agriculture and plant use at Tiwanaku: social and political implications. In A. Kolata (Ed.), *Tiwanaku and its hinterland: archaeological and paleoecological investigations of an Andean civilization* (Vol. 2, pp. 384–403). Urban and Rural Archaeology. Washington, D.C.: Smithsonian Institution Press.

Yarnell, R. A. (1974). Plant food and cultivation of the Salts Cavers. In P. J. Watson (Ed.), *Archaeology of the Mammoth Cave Area* (pp. 113–122). New York: Academic Press.

Chapter 4
Rethinking the Role of Wild Resources in Agriculturalist Societies: Archives from Rockshelter Cases of Northwestern Argentina

M. Alejandra Korstanje

Introduction

The concepts of wild and domestic in archaeology fundamentally influence the way we construct our knowledge about plants and societies. Often, when we are discussing wild or domestic resources we are not really following an accurate biological definition (Rindos 1984); we do not genetically check every specimen we find, nor do we even check the morphological attributes of them all (see Lema 2011). When archaeologists find domesticates in sites of agricultural societies we assume that they are indeed domesticated items.

Yet, this distinction has consequences for the ways archaeologists conceptualize societies, especially within evolutionary models that separate hunter–gatherers (wild plant users) from farmers (domesticated plant users) (see discussions in Gepts 2002; Lema 2008; Oyama 2002; Yacobaccio and Korstanje 2008). For example, in Northwest Argentina, a periodization scheme widely used until recent decades (González and Pérez 1971) characterized two principal periods based on whether or not the societies had acquired the technological features of ceramics and domesticated plants: *Período Precerámico* (Pre-ceramic Period) versus *Período Agroalfarero* (Agro-ceramic Period) (Korstanje et al. 2015). Such schemes become problematic, however, when faced with situations such as the importance of wild animal hunting during Inca times (Madero 1994) or the importance of *algarroba* (carob) gathering during colonial times (Arana 1999). Thus, greater effort is needed

For Dalmacia, Simonita, Isabel, Lorenza and Leocadio (†), wherever they are, they will recognize that I'm talking about their loved landscapes.

M.A. Korstanje (✉)
Instituto de Arqueología y Museo (FCN e IML, Universidad Nacional de Tucumán), Instituto Superior de Estudios Sociales (CONICET/UNT), San Martín 1545, 4000 San Miguel de Tucumán, Argentina
e-mail: alek@webmail.unt.edu.ar

M.P. Sayre and M.C. Bruno (eds.), *Social Perspectives on Ancient Lives from Paleoethnobotanical Data*, DOI 10.1007/978-3-319-52849-6_4

in order to deconstruct such schemes and their underlying assumptions about plants and their roles in people's lives.

Additionally, the study of wild or gathered vegetal resources is often relegated to a secondary place in archaeology as we tend to give more attention to the study of crops as *artifacts,* or plants created by humans (see Lema 2010). While there certainly are exceptions, this disparity can be seen in the many volumes dedicated to identifying and documenting domesticated plants (e.g., Cowan and Watson 1992; Iriarte and Vrydaghs 2007; Zeder et al. 2006). Moreover, when considering social dimensions of plant resources use we tend to focus on domesticates, particularly the food plants, even when wild species are present in the botanical inventory (unless they are hallucinogenic).

This disparity may not be intentional, but this is possibly due to the perception that there are social–cognitive patterns that operate through concrete actions to transform these domesticated plants, when in reality, cognitive patterns also operate on the selection, harvesting, use and modification—or *artefactualization*—of wild species as well. Perhaps it is because domesticated crops seem to have the imperial mark of invention and human labor that in some ways serve to dominate nature, thereby marking a special moment in our history as a species. Yet, a great deal of labor and collective work can go into obtaining wild resources for special occasions or as products of long distance exchange (Núñez and Dillehay 1979; Albeck 1994; Lazzari 2005). We might also ask, did people in the past separate this role of plants based on whether or not they were gathered or cultivars? Were wild and domestic also emic categories? Are wild and domestic emic categories today?

If we agree that the categories we use to construct our knowledge are analytical tools needed to sort and understand our cosmos, we should not expect that every category we build fits perfectly in all the *realities* we want to explain. They are just tools to start with, but we are allowed to change our tools as many times as we need: (1) to continue modeling a targeted past; (2) as our knowledge becomes more complex; and (3) if we learn more about what was previously unknown.

This seems to be the case with most of the traditional analytical binary categories in archaeology, inherited from Cartesian thought (Latour 2001): hunter–gatherers versus farmers–pastoralists or even simple versus complex societies. Such categories were useful for functionalist interpretations, but become uncomfortable when considered in light of other theoretical frameworks (Viveiros de Castro 1996; Ingold 2000; Descola 2012).

Ethnographic and ethnobotanical studies illuminate the fact that the traditional wild/domestic dichotomy may be forcing too narrow a view of subsistence practices onto archaeological contexts (Vickers 1989; O'Shea 1989; Sponsel 1989; Zimmerer 1996). With archaeological evidence of agriculture present, it is often more difficult to explore variability and alternative practices, including the possibility that people might possibly abandon agriculture and adopt hunting and gathering (Baleé 1994). Such studies also reveal how nonfarming activities can create modifications of the environment "including intensification of 'natural' patterns of distribution and long-term genotypic changes in plant and animal resources," (Ellen 2000: 198).

While *domesticated* may have been a powerful tool for explaining and understanding changes in transitional economies (for instance, as hunter and gatherers began cultivating), the *wild* remained an obscure category for those scholars involved in problems related to agricultural societies. Many theories of agricultural societies center on communally organized work (Raffino 1973), but how much work is invested in watering, harvesting, tilling, cropping, etc.? In Argentina, almost none of these categories were constructed around how much work is invested in collecting resources that sometimes are far away, that must be gathered, gained, reached, peeled, processed, stored during the trip and in the final destination. Is that a different kind of work? Don't these tasks also require labor organization rules?

I propose in this paper that labor investment and organization could be a key way to distinguish from a social point of view what we have previously classified as wild (gathered) or domestic (cultivated). Labor is always a precondition of any human economy, but there are additional dimensions that enrich labor (Korstanje 2005). One dimension worth considering is that plants are actors with agency and important roles in society, not only ecofacts to be merely used or eaten (Hastorf 1998). Plants are embedded in societies through activities such as gathering, planting, producing, preparing, consuming, exchanging and idolizing. Therefore, their role(s) in society are active not only as symbols, but often as symbiotic agents that nurture and care for others.

The Wild and Domestic in Argentine Rockshelters

An illuminating example of the differentiation between *wild* and *domestic* is in the archaeology of rockshelter occupations in Argentina, which are not usually associated with agriculturalists. In this paper, I examine rockshelter plant assemblages in Argentina as a means to consider the societies that used them and the plants found within them along a continuum for two main reasons: 1) the composition of rockshelter plant assemblages are much more coherent and complete as organic preservation is better in these contexts; 2) rockshelters as living spaces were often assumed to be part of the hunter and gatherer versus agriculturalist (or food producers) dichotomy, and it provides us a clear opportunity to break down this binary.

With some exceptions, rockshelters occupied during the agricultural period were not considered *domestic enough*, and researchers continuously described the vegetal resources found in these contexts as wild or domestic, but were clearly most interested in the crops (for a synthesis see Tarragó 1980). Moreover, in many cases, the gathered resources were not identified or even recorded.[1] On the contrary, when the context is supposed to be primarily *wild*—as the case of hunting and gathering

[1]See Appendix 1 for the Northwestern Argentinean cases.

societies—the *domesticated* plants gained more importance and were placed in new, ambiguous categories such as *foreign* or *exchanged* plants. While some ethnoarchaeological cases support the idea that rockshelters within agricultural societies are transit sites (see Goñi 1995), and thus plant remains from different environments are probably to be expected, it is interesting to see that this obsession with *foreign* resources does not appear in relation to open-air sites. In these cases, the research questions tended to be oriented toward broader social problems such as territoriality, status and power, redistribution of goods, agricultural intensification, and so on.

It is also necessary to consider that rockshelters are not the only available living spaces for farmers. The significance of these spaces can change in the social landscape, and thus for farmers rockshelters might constitute a part of a productive space (as corrals, for instance). Therefore, rockshelters might be showing us that the diversity of resources used and needed by people—including for food availability, artifact raw materials, ritual offerings, etc.—come from a broad range of places. They demonstrate many different strategies of exchange, production and reproduction, and bring out different labor situations and conditions. *Wild* is not fully wild anymore (if it ever was). The items we formerly identified as wild, whether animal or vegetal, have the same requirements as the ones that people grew—interaction, work, labor organization, negotiation, and family networks.

Rockshelters in Northwestern Argentina (NWA)

The archaeology of Northwestern Argentina (Fig. 4.1) has focused on many concepts that are involved in the present arguments: economies, interaction, domestication, hunter and gatherer transitions to agriculture, labor, and social organization of early societies. Chronologies proposed for the different regional historical periods were established around strong distinctions between food-producing versus foraging societies—even when other key aspects of societies might have been involved, such as mobility, technology, architectural organization, or modes of production (González 1980; Núñez Regueiro 1975). In this chapter I will be drawing from a wide range of studies conducted in rockshelters of this region of Argentina (Aguerre et al. 1973; Aschero 1980; Aschero and Yacobaccio 1998–1999; Castro and Tarragó 1992; Fernández 1988–1989; Fernández Distel 1974; García 1988/1989, 1998; Lavallée et al. 1997; Rodríguez 1997, 1999a, b, c, 2000, 2004; Rodríguez and Rúgolo de Agrasar 1999). From this starting point, local archaeologists distinguished agricultural/pastoral societies—those that produced food—strongly stating that the main subsistence resources were no longer *wild* but *domestic*. In that sense, animals, food, wild plant resources, technology, and even social relationships became *less wild*.

The time of initial agricultural development is known in local chronologies as the Formative Period (ca. 2000 BC–1000 AD), and agriculture plays a key role in its definition. Agriculture is assumed by many Argentine archaeologists to be

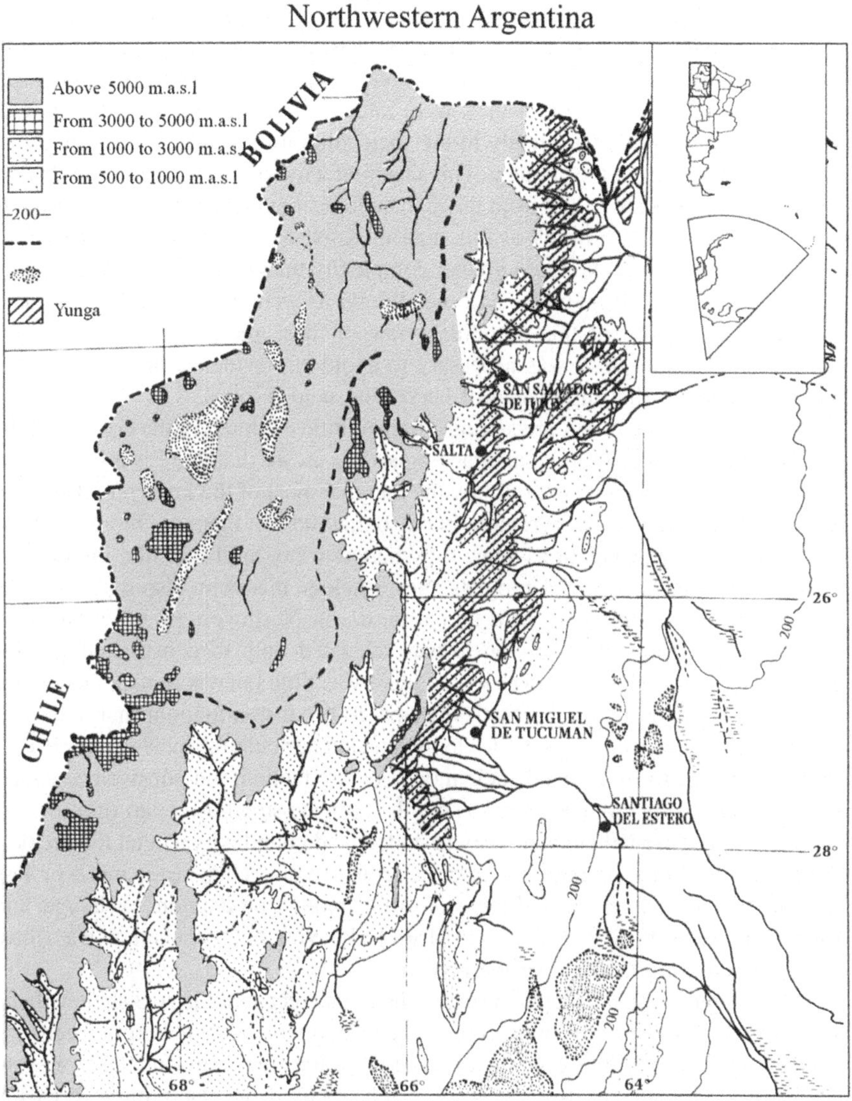

Fig. 4.1 Study area of Northwest Argentina

associated with a sedentary lifestyle, but with a different degree of population mobility, according to different landscapes, within the region and according to the ideas of the permanence or precariousness of their homes (Raffino 1973; Olivera 1991; Albeck 2000).

NWA has been considered the agricultural zone *par excellence* of the pre-Hispanic past in Argentina. It is home to a vast expanse of pre-Hispanic fields, whose now abandoned land and structures were once full of life and verdant crops. Archaeologists have argued that they were intensively exploited and that agricultural production contributed greatly to the sociocultural matrix of this region, more so than in other areas of the country. This vision of a more *agriculturalist* area is not an image that can be sustained from the archaeological evidence, since pre-Hispanic agriculture exceeded the limits of this region (Korstanje and Quesada 2010). The bias that this region was home to greater complexity than others is based in part on its architecture, ceramics, textiles and metallurgy, as well as the fact that the Incas conquered it. Because of this particular view of these people as more *civilized*, *complex* and *Incanaized*, it was necessary to avoid any evidence that contradicted this vision, such as the importance of harvesting wild resources.

Rockshelters in the NWA are settings that are little explored in the context of the agricultural societies. Argentine archaeologists are aware that there is a bias in this selection of sites, produced not only by the fact that most of the known rockshelters are in the Puna (high Andean) region, but also that most of them are more carefully studied for Archaic occupations since rockshelters were, until recently, the favorite target of hunter/gatherer archaeologists. Nevertheless, this point also reinforces the construction of the categories *wild* and *domestic* in Northwestern Argentina's archaeology. For many years, *domestic* was relegated only to constructed spheres such as household, villages, and cities, so there was little interest in rockshelters for scholars studying agricultural societies (Lema 2009). It was thought that there was no need to go beyond these built sites to understand agricultural societies: they were sufficient enough to understand their modes of production and subsistence strategies. Although I agree that economy is not the ultimate structuring part of a society, from my point of view there is no way to understand complexity, social hierarchies, politics and ideology or more abstract structures without having a more or less accurate idea of the three basic dimensions of economy: production, exchange, and consumption (Costin 1991; Marx 1946 [1867]). Leaving out rockshelters was, thus, leaving out an important part of the economy.

In this chapter, I include a panorama of the plants recovered in rockshelters from Northwestern Argentina up until 2003, when this theoretical dichotomy between wild and domestic predominated the discussions in the area. Unfortunately, relevant publications were scarce, dispersed, and rarely focused on plants until about ten years ago. Since then, new research with more robust recovery strategies and analytical methodologies has been conducted and published; changing the axis of the discussion in many ways, but the pattern shown in this work has not changed consistently for those more interested in plant data itself. Therefore, examining these data presents an opportunity to engage in a reflective consideration regarding our own professional practices and changing ideas.

As I will try to show, gathered wild resources in agricultural societies in Argentina play an important role, as part of social practice, that expands beyond the wild sphere. The gathering of wild resources is a social practice charged with strong economic importance and social meaning, and these dimensions can be ascertained by exploring the labor organization involved in their procurement (Lema 2009; Quiroga 1999, 2010).

Rockshelters and Their Plant Remains

I begin with a broad exploration of published and unpublished information on rockshelters before discussing whether these resources were called wild or domestic by archaeologists. Rockshelters provide a good case study because of their excellent organic preservation conditions with respect to the archaeobotanical (macro and microbotanical) and zooarchaeological records, which are both well preserved and recovered. These contexts can also be readily compared. I focus primarily here on the layers that are dated and interpreted as *agriculturally productive moments*, even though for comparison I will include some hunter and gatherer occupation layers.

Appendix 1 provides a summary of the plant record for rockshelter sites in Northwestern Argentina that have been excavated up to 2003 (whether published or not, with botanical records or not). The plant remains are organized into three categories: organic artifacts, edible, and nonedible plants. The Appendix also includes information on the recovery methods used. Table 4.1 summarizes the patterns in plant remains from hunter–gatherer contexts in the rockshelters and Table 4.2 summarizes the patterns from agriculturalist contexts. Table 4.3 provide some additional information for interpreting the role of the origin and role of the wild species in the rockshelters. Table 4.3 synthesizes information on the modern-day uses of local wild species in dry and semi-dry environments of Argentina including food in its different forms, medicine, ritual, artifact production, etc.

Based on Table 4.1 and the appendix data, we are able to discuss some general patterns in hunter and gatherer contexts, and to orient us to what species have traditionally been considered *wild*.

As expected, the preagriculture occupation layers show that there is a predominance of wild resources from grasses to woods. However, a domesticated crop, the bottle gourd (*Lagenaria siceraria*) is present. Although it is not in the earliest contexts, it shows up around AD 1000 when hunter–gatherers were still common in the area. It was likely used as a container but is also edible when unripe. Gourds grow in the lower valleys, which are not really very far from the rockshelter sites. However, they were likely not produced by the people using them in the rockshelters, and have been considered by archaeologists to be an exchange good.

Table 4.1 Vegetation and artifacts (presence/absence) in preagriculture occupations—ca. 7700–4080 BP (see Appendix 1 for species and details)

SITE	ARTIFACTS		NON-EDIBLE PLANTS				EDIBLE PLANTS								
			Wood	Grasses	Psychotropic /Stimulants	Containers	Cactus fruits	Tree fruits, pods	Grasses	Tubers/ roots	Legumes	Vegetables	Pseudo-cereals	Spices	Other
Inca Cueva C4	wild		wild	wild			wild			wild	wild				
Huachichocana III (Level E3)	wild		wild	wild			wild	wild			wild			wild	X
León Huasi				wild				wild							
Inca Cueva C7	wild	dom	wild	wild	wild	dom	wild	wild							X
Alero de las Circunferencias															
Cueva de Cristobal															
Tomayoc															
Quebrada Seca III	wild		wild	wild											
Punta de la Peña 4	wild	dom	wild	wild		dom									
Salamanca	wild		wild	wild											
Puente del Diablo								wild			X	X			X

Another observation of this assemblage is that there is greater variability in wild species used as artifacts and nonedible plants than edible plants. Edible plants are restricted to cactus fruits, tubers and *Prosopis* spp. pods. Several of the species, including cactus, *Prosopis* spp., wild legumes and spice species are not local to the puna where the Rockshelters are located. Thus, this suggests an additional amount of work and/or exchange required for the collection of these plants.

Wood and grasses are always included in the wild sphere. Some of them, including *Polylepis tormentella* (queñoa), *Azorella compacta* (yareta), *Chusquea* sp. (cañas/canes), derive from far distances, which imply a different attitude than just *gathering* them locally. They are specially selected and labor is invested in carrying them, both physically and socially. While these species could be used for fuel, we find them in the rockshelters used for artifact production.

Finally, archaeologists at the site of Inca Cueva 7 recovered seeds of *Anadentanthera colubrina* or cebil. This wild resource is known to be a psychotropic plant. It would have been obtained from great distances (Yunga area, see Fig. 4.1) and used for ritual practices (Constantino 1998; Pérez Gollán and Gordillo 1994; Pochettino et al. 1999).

Based on Table 4.2 and the appendix data, encompassing the agricultural occupations, we observe that some general trends encountered in hunter–gatherer contexts are repeated, perhaps unexpectedly. There is a similar range of wild species utilized and relatively fewer wild edible species than artifacts and nonedibles. Only two trends are significantly different from the hunter–gatherer contexts. First, as expected, the agricultural occupation layers show both the presence of wild and domesticated resources. Second, there is actually *greater* variability in wild species utilized by the agricultural groups.

The presence of crops including tubers and the pseudocereals quinoa and amaranth may be explained by the fact that these items were broadly produced in these later periods and as such were regularly entering into the system as common goods. However, many of the crops, notably maize, beans, and peanuts do not grow in the puna environments where the rockshelters are located. Thus, it is possible that some form of the earlier labor and exchange strategies would have continued to be involved in their procurement. The crops may or may not have been produced by the people that were consuming them at these shelters.

Although the majority of the edible plants are crops, wild edible plants still appear to play an important role here, and include many of the species encountered in earlier periods such as cactus fruits and *Prosopis* spp. This is significant given that it is quite likely that the wild/gathered items listed in these reports are actually incomplete because early archaeologists were more interested in crops from the Formative period on. The important role of a diversity of wild plant species is observable in the region today as evidenced by the large number of wild plants local people distinguish today (see Table 4.3). According to these sources, most of these plants are used for construction materials and fuel, rather than food.

Table 4.2 Vegetation and artifacts (presence/absence) in post-agriculture occupations—ca. 950 BC to AD 1300 (see Appendix 1 for species or details)

SITE	ARTIFACTS	NON EDIBLE PLANTS				EDIBLE PLANTS								
		Wood	Grasses	Psychotropic /Stimulants	Containers	Cactus fruits	Tree fruits, pods	Grasses	Tubers /roots	Legumes/ Oily	Vegetables	Pseudo-cereals	Spices	Other
Inca Cueva C1	dom				dom			dom						
Inca Cueva C5		wild						dom	wild					
Inca Cueva C1	wild	wild	X											wild
Huachichoc. III (Levels E1, D)	wild dom	wild	wild		dom	wild	wild	dom	dom	dom				X
Huachichocana V	wild			dom				dom	dom	dom	dom			wild
Quebrada Seca III (Levels 1, 2a)	wild	wild	wild											
Real Grande 3	wild dom		X		dom									
Cacao I														
Punta de la Peña 4	wild dom	wild	wild		dom		wild	dom				dom		
Punta de la Peña 11	wild	wild	wild											
Los Viscos	wild dom	wild	wild		dom	wild	wild	dom	X	wild dom	dom	dom		X
Cueva Pintada														
Quebrada de los corrales								X	X			dom		X
Cueva El Litro	wild dom						wild	dom		dom	dom	dom		X

Table 4.3 Number of wild plants recorded in the present days used for different purposes by people in different environments

Environment	Number of wild species used in the present	Cited in
Dry Puna	31	Garcia (1998)
Dry Puna	28	Yacobaccio (1990)
Salt Puna	40	Rodríguez (1998)
High Valleys	30	Babot (1999, 2001)

Focusing on Farmers' Livelihoods, What Is the Role of Gathered Resources?

Even within social archaeology, we still think it is fundamental to know the context of production in order to understand social change, power relationships and sacred rituals in this part of the Andes (Costin 1991; Hastorf 1998, 2003; Jennings and Bowser 2008; Sayre 2014). Regarding early pre-Hispanic agricultural systems, most of the research has been conducted in the valleys, where the majority of the Andean crops could be grown, and counts on the recognition of cultivated species or carbonized seeds of some type to be found in domestic contexts. Unfortunately, we do not have information about the first phases of experimentation and initial development of agriculture in Northwestern Argentina (Lema 2008, 2009). Agriculture was already an established system at our earliest sites. Thus, with regard to the *first* farmers, we do not really know how they *learned* to farm, how much they relied upon animal (camelid) husbandry and pastoral practices, or how they included those practices in their everyday life and cosmology. This is not essential, however, when the point of this chapter is not to reconsider these widely debated binary terms and concepts, but rather to consider the role of wild and domesticated plants along a *continuum* of human subsistence systems. Thus, we may ask instead: how different are the farmers' practices and uses of gathered, *wild* resources than those of the earlier hunter–gatherers who had started to implement horticulture?

In some ways, we might even consider the greater importance of wild species to small-scale farming societies, who are frequently vulnerable to severe fluctuations in agricultural production. "For such systems to approximate self-sufficiency, it is crucial that other, highly productive food resources be available, whose structure of interannual variability is largely independent of that governing agriculture," (O'Shea 1989: 57). In this sense, it is important to maintain or even increase access to the gathered resources that have been consumed prior to the presence of crops, such as the rich tree seeds and fruits.

In order to explore the importance of the wild sphere in the life of people who we characterize as agriculturalists, I explore a specific example from my own research in the El Bolsón valley of Northwest Argentina (Fig. 4.1). Here, we included rockshelter occupations as a part of the broader study of ancient agricultural landscapes and societies—and quickly we realized that it is necessary to go beyond the customary wild/domestic categories to understand the plant assemblages we were encountering here (Korstanje and Würschmidt 1999). The Los Viscos shelter record (Table 4.2 and Appendix 1), for instance, showed that the presence of wild resources was quite important comprising 75% of the seed sample, even in a time period when maize, squashes, pseudocereals and beans were regularly present. So, we shifted our question to ask *how wild* were these gathered resources? Were they regularly gathered and did they grow near the settlements? Perhaps these plants also moved with the human populations and had strong caloric, economic and ideological value, as is the case with the algarrobo (*Prosopis alba et nigra*), a wild species that is intensively managed for fuel and food by agriculturalists (Quiroga 1999; Arana 1999).

For other species, such as the psychotropic seed producing trees of cebil or *vilca* (*Anandentanthera colubrina* var. *cebil*), perhaps they were not planted by humans near their houses, nor formed part of their regular diet, but were more represented in social imagery than the edible cultivated crops. Such considerations reveal the importance of focusing more attention on the identification of archaeological *wild* resources, even if they occur in the same contexts as the supposedly more important staple crops, such as maize.

Rockshelters and their excellent preservation reveal the great diversity of resources used and needed by people, for food, artifact raw materials, ritual offerings, etc. They also show that the plants people used came from a broad range of ecological settings, illuminating many different strategies of exchange, production and reproduction, which bring out different labor situations and conditions. Maybe the difference between wild and domestic is not quite real, at least in the consumption sphere or in the meaning sphere, but obviously in the *productive* sphere.[2] The difference is that when people *produce* they involve new programmatic activities that have to be followed if success is to be achieved- which include close timetables and the right labor organization. These activities imply clear power rules and strengthened power relationships. Thus, *wild* is not fully wild anymore (if it ever was). The plants we formerly identified as wild have as many requirements of interaction, work, labor organization, negotiation and familial networks as the ones that people grow.

Conclusions

> Most of those who have sought to understand human culture have been prepared to accept subsistence practices as basically unproblematic, requiring for their analysis no more than simple typologies and a bit of common sense, (Ellen 2000: 197).

[2]The obviousness is only apparent, as in these "productive" spheres usually involved hunting and gathering practices.

Subsistence strategies connect to many areas of theoretical inquiry. In current archaeological research there is an important stream that tends to focus on what has been formerly considered *superstructural* problems, such as ideology, sex, gender, power, etc. I agree with Ellen (2000) that there is still a lot to understand about modes of subsistence, and there are still many themes that have been little discussed in ongoing research and through the literature.

The point of this theoretical exercise was to explore how useful such categories are in the construction of archeological knowledge. While biologists agree on the concept of *domestication*, sharp-edged distinctions of wild and domestic categories were also useful for early archaeologists (see a discussion in Hayden 1995). This is certainly true, especially for archaeologists studying transitional moments where people changed their livelihood from a predominantly hunter and gatherer focus to a new agropastoral one. But this distinction is not open enough for understanding those societies that already have the agricultural and/or pastoral skills embodied in their daily life. Even if "there appears to be a *conceptual gap* between the role that plants and plant knowledge played in the past, and the level of research interest and commitment to study of paleoethnobotanical data within archaeology" (Archer et al. 2000: 33), the importance of plants in resolving this conceptual gap makes this gap unacceptable in current archaeology.

Labor investment and organization could be a key place from which to distinguish, from a social point of view, what we have previously classified as wild (gathered) or domestic (cultivated). Even if labor is always a precondition of any human economy (Marx 1946; McGuire 1992), there are other dimensions that enrich it: whether it is individual labor, group solidarity labor (reciprocity), consented labor offer (as in a redistribution situation) and labor appropriation by force (slavery) (Morrison 1995; Korstanje 2005). Accurate methodologies to approach such discussions are not easy to find. As with other archaeological problems in exchange studies, labor indicators lead archaeologists to stumble into thinking that the biggest things need a bigger labor investment, which is simply an assumption. There is much to explore about that theme (Korstanje 2010), but the good news is that labor is a concept closely attached to economy, and archaeologists have already done a good job in studying that successfully.

Acknowledgements This paper was first written when I attended the Paleoethnobotany Graduate Course with Christine Hastorf, and as a part of a Fulbright Senior Research Scholarship at the Paleoethnobotany Laboratory, University of California at Berkeley (2000–2001). It was originally presented at the SAA meeting at New Orleans (2001) in a session organized by Emily Dean and Kathy Twiss exploring "wild and domestic" but has been enriched, updated and modified for this book. Although new research and publications have emerged in Northwestern Argentina since 2003 the pattern shown in this study has not changed consistently. A special thanks to Maria Bruno and Matt Sayre for inviting me to participate in this book, and of course, to Christine Hastorf, my dear teacher and now my friend.

Appendix 1

Botanical Remains from Rockshelter sites in the Puna and Valleys of Northwestern Argentina recorded up to 2003. Included are names of the sites, date ranges as provided in the original publication, plants encountered and other significant finds.

Codes for recovery information:

I. Sites without botanical record
I-1. Even if accurate recovering techniques were used
I-2. Recovery techniques were not explicit
I-3. Botanical remains are not mentioned
II. Sites with botanical record
II-1. Recovery techniques are not explicit
II-2. Micro and micro analyses are involved.

(A) **PUNA**

- ***Inca Cueva C1*** (Azul Pampa, Jujuy). [II.1]

Chronology: supposedly ca. 1000 A.D. (Late Period)

Artifacts: *Lagenaria* sp. (calabaza matera/bottle gourd)

Edible plants:
Grasses: *Zea mays* (maíz/maize)

- ***Inca Cueva C4*** (Azul Pampa, Jujuy) [II.1]

Chronology: 9230 ± 70 BP

Artifacts: Basketry. Wild animals (cervidae) and mollusks (*Strophocheilus oblongus*), feathers of jungle birds

Nonedible plants:
Wood: *Polylepis tormentella* (queñoa), *Azorella compacta* (yareta), *Trichocereus tarijensis* (cardón), *Prosopis ferox* (churqui), *Chusquea* sp. (cañas/canes)
Grasses: *Cortaderia*sp, *Festuca* sp, *Stipa* sp., *Bromus* sp.

Edible plants:
Tubers and roots: *Hypsocharis* sp (soldaque, only wild varieties)
Cactus fruits: *Opuntia* sp. (airampo)
Legumes: *Lupinus* sp.

- ***Inca Cueva C5*** (Azul Pampa, Jujuy) [II.1]

Chronology: 1200 ± 60 BP to 780 ± 100 BP

Non-edible plants:
Wood: unidentified

Edible plants:
Grasses: *Zea mays* (maíz/maize)
Tubers and roots: wild fresh potatoes

- ***Inca Cueva C6*** (Azul Pampa, Jujuy) [1.3]

- ***Inca Cueva C7*** (Azul Pampa, Jujuy) [II.1]

Chronology: 4080 ± 80 BP (2130 BC)

Artifacts: hard wood, vegetal fibers and cane. Bottle gourds: *Lagenaria siceraria.* Also artifacts done on wild animal bones (Camelidae, *Felix concolor*, *Tapirus terretris*), or mollusks (*Oliva* sp.)

Nonedible plants:
Wood: unidentified, *Juncus* spp.
Grasses: *Cortaderia* sp, *Oenothera* sp., *Chusquea lorentziana* and others unidentified
Hallucinogenic: *Anadentanthera macrocarpa* (cebil)
Edible plants:
Cactus fruits: *Airampo* sp. and *Trichocereus pasacana* (pasacana)
Tree fruits, pods: *Prosopis ferox* (churqui), *Prosopis alba (*algarrobo blanco),
Unidentified seeds
Unidentified fruits

- ***Inca Cueva shelter 1*** (Azul Pampa, Jujuy) [II.1]

Chronology: 2900 ± 70 BP

Artifacts: wood
Nonedible plants:
Wood: *Trichocereus pasacana* and unidentified
Grasses: unidentified
Edible plants: *Juelia* sp. (mushroom, possible edible)

- ***Inca Cueva shelter 3*** (Azul Pampa, Jujuy) [I-3]

Chronology: supposed ca. 1000 A.D. (Later Period)

- ***Huachichocana III*** (Tumbaya, Jujuy) [II.1]

Level E3. Chronology: 9620 ± 130 BP (7670 BC), 8670 ± 550 BP (6720 BC), 8930 ± 300 BP (6980 BC)

Artifacts: Wood, canes, vegetal fibers, (*Sporobolus rigens*). Feather, wool.

Non-edible plants:
Wood: *Trichocereus* sp., *Juncus* sp.
Grasses: *Stipa eristachya*, *Sporobulus rigens*, (pasto espuro), *Tipha* sp. (totora), *Cortaderia* sp.
Other: *Tillandsia* sp. (clavel del aire), chiragua

Edible plants[3]:
Cactus fruits: *Opuntia* sp., *Trichocereus pasacana*
Tree fruits, pods: *Prosopi snigra*
Legumes: *Phaseouls vulgaris* (poroto/bean, wild variety)
Spices: *Capsicuum baccatumo C. chacoense* (ají/chili, possibly a wild variety)
Other: *Passiflora* sp.

Level E1.[4] Chronology: AD 730

Artifacts: *Lagenaria* sp.

Edible plants:
Maní/peanuts: *Arachis hypogaea*
Cactus fruits: *Trichocereus pasacana*
Tree fruits, pods: *Prosopis alba*
Grasses: *Zea mays*
Tubers and roots: *Oxalis tuberosa* (oca)

Level D. Chronology: AD 1100–1300

Artifacts: vegetal fibers.

Non-edible plants:
Wood: *Trichocereus pasacana* and other unidentified
Poaceae: *Tipha* sp.
Other: *Abromieitiella* sp. (amara), chiragua

Edible plants:
Grasses: *Zea mays* (maíz)
Unidentified seeds with evidence of chewing

Level C.[5] Chronology: XV–XVI centuries

Artifacts: *Lagenaria* sp., wood, Cactus, vegetal fiber and animal bones

Non-edible plants:
Wood: *Trichocereus pasacana*
Grasses: *Tipha* sp., *Cortaderia* sp., *Sporobulus rigens*,

---

[3]From this level there are remains of maize cobs, but from non-ancient varieties. On the other hand a C14 dated on the cob itself produced an age of 1560 ± 160 AP (Fernández Distel 1980 quoted in Yacobaccio 1990: 96). Attributing an early age to maize is still a common error in most of the general manuals about Northwestern Argentinean archaeology and its earliest crops (specially the international ones); it's important to stress here that this is not correct.

[4]This information is provided by Tarragó (1980), but is not present in Fernández Distel (1974). It could be part of a personal communication or of "… unpublished notes given by the author" (Tarragó 1980: 201). Here I follow Tarragó as it a more recent publication.

[5]Again, there is no agreement among Tarragó information (1980) and Fernández Distel (1974).

Edible plants:
Cactus fruits: *Trichocereus pasacana*
Tree fruits, pods: *Juglans australis* (nuez criolla/local nut), *Prosopis alba et nigra*
Grasses: *Zea mays*
Tubers and roots: *Canna edulis* (achira)
Vegetables: *Cucurbita* sp. (zapallo/squash)
Legumes: *Phaseolus* sp. (possible *lunatus*)
Spices: *Capsicum* sp.
Oleaginous: *Arachis hypogaea*
Unidentified seeds and leaves with evidence of chewing

- ***Huachichocana V*** (Tumbaya, Jujuy) [II.1]

Chronology: Early Formative (ca. 600 b.C to 100 d.C)

Artifacts: wood, vegetal fiber, and animal bone
Non-edible plants:
Stimulants “coca”
Edible plants:
Grasses: *Zea mays*
Tubers and roots: “achira”
Vegetables: “calabaza”
Oleaginous: *Arachis hypogaea*
Unidentified “wild” seeds

- ***León Huasi*** (Jujuy)

Chronology: (we do not have access to the original publication)

Non edible plants:
Poaceae: *Stipa eriostachya.*, *Sporobolus rigens.*

Edible plants[6]:
Tree fruits, pods: *Prosopis nigra*

- ***Tomayoc*** (Sierra del Aguilar, Jujuy) [I.3]

Chronology: 4250 ± 50 BP to 550 ± 50 BP (1435 d.C)

- ***Cueva de Cristobal*** (Humahuaca, Jujuy) [I.3]

Chronology: 2860 ± 160 AP
No data (1.2)

- ***Alero de las circunferencias*** (Humahuaca, Jujuy)

Chronology:
Not published info found

[6]Ancient remains of maize are discussed in this publication (Fernández Distel 1980 quoted in Yacobaccio 1990: 96).

- ***Quebrada Seca III*** (Antofagasta de la Sierra, Catamarca) [II.1]

Levels 1 to 2a. Chronology: 2480 ± 60 BP (AD 530).

Artifacts: Basketry (*Cortaderia* sp.)

Non-edible plants:
Wood: *Salix humboldtiana*, *Adesmia horrida*, *Fabiana punensis*, *Senecio santelicis*, Asteraceae spp.
Poaceae: *Deyeuxia eminens*.

Level 2b1 to 2b25.[7] Chronology: 5400 ± 90 BP (AD 3460) to 9410 ± 120 BP.

Artifacts: *Parastrephia quadangularis*, *Adesmia horrida*, *Atriplex* sp., *Prosopis torquata*, *Salix humboldtiana*, *Acrocomia totai*, *Chusquea lorentziana*, *Rhipidocladum neumannii*, *Deyeuxia eminens*, *Cortaderia speciosa*.

Non-edible plants:
Wood: *Adesmia horrida*, *Fabiana bryoides*, *Fabiana punensis*, *Parastrphia lucida*, *Parastrephia quadangularis*, *Baccharis incarum*, *Senecio santelicis*, *Sisymbrium philippeanum*
Grasses: *Deyeuxia eminens*, *Festuca weberbausi*, *Festuca ortophylla*, *Festuca chrysophylla*, *Stipa* sp., *Puccinellia frigida*.
Other (vegetative or reproductive isolated organs): Fabaceae spp. (*Arachis monticola*), Asteraceae spp., Solanaceae spp.

- ***Real Grande 3*** (Antofagasta de la Sierra, Catamarca) [II.1]

Chronology: 770 ± 60 to 680 ± 79 BP

Artifacts: Basketry (*Juncus* sp.), *Lagenaria siceraria*

Non-edible plants:
Grasses: unidentified

- ***Cacao I*** (Antofagasta de la Sierra, Catamarca)

No published information found

- ***Salamanca 1*** (Antofagasta de la Sierra, Catamarca) [II]

Chronology: 7410 ± 100 BP

Artifacts: *Chusquea lorentiziana* Griseb.;*Trichocereus pasacana* (Web) Britton et Rose

Non-edible plants:
Wood: *Adesmiahorrida* Gillies ex Hook & Arn.; *Baccahris incarum* Wedd; *Fabiana bryoides* Phil.
Grasses: *Deyeuxia eminens* J. Presl. var. *fulva* (Griseb.) Rúgolo; *Festuca* sp.

[7]This is one of the best contexts studied from a paleoethnobotanical perspective in the Argentinean Puna. For a complete reference see Rodríguez (2000).

Edible plants:
Legumes: *Hoffmanseggiaeremophila* Phil.

- ***Punta de la Peña 4*** (Antofagasta de la Sierra, Catamarca) [II]

Chronology: 4060 ± 90 to 3870 ± 90 BP and 3250 ± 50 to 460 ± 79 BP (not specifically differentiated in the bibliography)

Artifacts: *Lagenaria siceraria* Ser.; *Trichocereus pasacana; Deyeuxia eminens*; *Deyeuxia deserticola*

Non-edible plants:
Wood: *Adesmia horrida*; *Fabiana bryoides*; *Fabiana punensis* S.C. Arroyo; *Neospartone phedroides; Acantholippia deserticola; Parastrephia quadrangularis* (Meyen) Cabrera.
Poaceae: *Deyeuxia eminens*. var. *fulva; Deyeuxia eminens*; *Pennisetum chilense* (Desc) B.D. Jackson ex R.E.

Edible plants:
Tree fruits, pods: *Prosopis* sp.
Grasses: *Zea mays* Linn.
Pseudocereals: *Chenopodium quinoa* Wild

- ***Punta de la Peña 11*** (Bebé de la Peña- Antofagasta de la Sierra, Catamarca) [II]

Chronology: 3210 ± 50 to 3630 ± 150 BP

Artifacts: Ropes: *Acrocomia* sp.

Non-edible plants:
Wood: *Adesmia horrida; Parastrphia quadrangularis; Neosparton ephedroides; Acantholippia deserticola.*
Grasses: *Deyeuxia eminens; Deyeuxia eminens* var. *fulva*

(B) **VALLEYS**

- **Puente del diablo** (*Cachi*, *Salta*) [II.1]

Chronology: supposed Archaic (ca. 10.000–3000 BP)
Edible plants:
Tree fruits, pods: *Prosopis* sp. and undetermined "wild" species
Vegetables: Cucurbitaceae seeds
Legumes: possible *Phaseolus* sp. bean

- ***Los Viscos*** (El Bolsón, Catamarca) [II.2]

Chronology: 2270 ± 230 BP

Artifacts: wood, vegetal fibers, *Lagenaria siceraria*, feathers and animal bone

Non-edible plants:
Wood: *Trichocereus pasacana*, *Acacia visco*, *Abromeitielia* sp. and unidentified
Grasses: *Cortaderia* sp. and unidentified

Edible plants:
Cactus fruits: *Trichocereus pasacana*, *Opuntia* sp.
Tree fruits, pods: *Geoffroea* sp., *Prosopis* sp.
Grasses. *Zea mays*
Tubers and roots: unidentified
Legumes: *Phaseolus* sp. and wild legumes
Vegetables: *Cucurbita* sp.
Pseudocereals: *Chenopodium quinoa*, *Amaranthu* sp.

Microremains (starches and phytoliths): *Zea mays*, *Cucurbita* sp., *Chenopodium quinoa*, *Amaranthus* sp.

- ***Cueva Pintada*** (El Bolsón, Catamarca) [II.1]

Chronology: 1790 ± 80 BP

Non edible plants:
Wood: unidentified
Grasses: unidentified

- ***Quebrada de los corrales*** (El Infiernillo, Tucumán) [II.2]

Chronology: formative (ca. 600 b.C to 900 d.C)

Microremains (starches): *Zea mays*; pseudo cereals: *Chenopodium* or *Amaranthus* sp.

- ***Cueva El Litro*** (Pampa Grande, Salta) [II.2]

Chronology: supposed ca. 700 A.D.
Artifacts: wood, vegetal fibers, *Lagenaria siceraria*.
Edible plants:
Tree fruits, pods: *Geoffroea decorticans*, *Prosopis alba o nigra*
Grasses: *Zea mays*
Legumes: *Phaseolus vulgaris*, *Phaseolus lunatus*
Vegetables: *Cucurbita maxima*, *y Cucurbita* sp. (*C. andreana?*)
Pseudocereals: *Chenopodium quinoa*, *Amaranthus caudatus*,
Other: unidentified seeds and fibers
Microremains (pollen): *Zea mays*, *Chenopodium* sp./*Amaranthus* sp.

References

Aguerre, A., Fernández Distel, A., Aschero, C. (1973). Hallazgo de un sitio acerámico en la Quebrada de Inca Cueva (Jujuy). *Relaciones N.S.*, *VII*, 197–235. (Buenos Aires: Sociedad Argentina de Antropología).

Albeck, M. E. (1994). La Quebrada de Humahuaca en el intercambio prehispánico. In M. Albeck (Ed.), *Taller "De Costa a Selva". Producción e Intercambio entre los Pueblos Agroalfareros de los Andes Centro Sur* (pp. 117–127). Jujuy: Instituto Interdisciplinario Tilcara (UBA).

Albeck, M. E. (2000). La vida agraria en los Andes del sur. In M. Tarragó (Ed.), *Nueva historia argentina. Los pueblos originarios y la Conquista* (pp. 187–228). Buenos Aires: Editorial Sudamericana.

Arana, M. M. (1999). El tiempo de la algarroba. In C. Aschero, M. A. Korstanje, & P. Vuoto (Eds.), *En los tres reinos: prácticas de recolección en el cono sur de América* (pp. 197–204). Tucumán: Instituto de Arqueología y Museo (UNT).

Archer, S., Hastorf, C., Coil, J., Dean, E., Goddard, R., Near, J., et al. (2000). Paleoethnobotany and archeology 2000. The state of Paleoethnobotany in the discipline. *SAA Bulletin, 18*(3), 33–38.

Aschero, C. (1980). Comentarios acerca de un fechado radiocarbónico del sitio Inca Cueva-4 (Departamento Humahuaca, Jujuy, Argentina). *Relaciones N.S, XIV*(1), 165–168. (Buenos Aires: Sociedad Argentina de Antropología).

Aschero, C., & Yacobaccio, H. (1998–1999). 20 Años después: Inca Cueva 7 reinterpretado. *Cuadernos,* (18), 7–18. (Buenos Aires: Instituto Nacional de Antropología y Pensamiento Latinoamericano).

Babot, M. D. P. (1999). Recolectar para moler. Casos actuales de interés arqueológico en el Noroeste argentino. In C. Aschero, M. A. Korstanje, & P. Vuoto (Eds), *En los tres reinos: prácticas de recolección en el cono sur de América* (pp. 161–170). Tucumán: Instituto de Arqueología y Museo (UNT).

Babot, M. D. P. (2001). La molienda de vegetales almidonosos en el Noroeste argentino prehispánico. XI Simp. Arg. de Paleobotánica y Palinología. *Publicación especial,* (8), 59–64. (Buenos Aires: Asociación Paleontológica Argentina).

Baleé, W. (1994). *Footprints of the forest, Ka'apor ethnobotany—The historical ecology of plant utilization by an amazonian people.* New York: Columbia University Press.

Castro, V., & Tarrago, M. (1992). Los inicios de la producción de alimentos en el Cono Sur de América. *Revista de Arqueología Americana* (6), 91–124. México: Instituto Panamericano de Geografía e Historia.

Constantino, M. T., (1998). Psychoactive substances in the archaeology of Northern Chile and NW Argentina a comparative review of the evidence. *Chungará, 30*(1), 49–63.

Costin, C. L. (1991). Craft specialization: issues in defining, documenting and explaining the organization of production. In M. Schiffer (Ed.), *Archaeological method and theory* (Vol. 3, pp. 1–56). Springer.

Cowan, W., & Watson, P. J. (Eds.). (1992). *The origins of agriculture an international perspective.* Washington D.C.: Smithsonian Institution Press.

Descola, P. (2012). *Más allá de naturaleza y cultura.* Buenos Aires: Amorrortu eds.

Ellen, R. (2000). Modes of subsistence: Hunting and gathering to agriculture and pastoralism. In T. Ingold (Ed.), *Companion encyclopedia of anthropology* (pp. 197–225). Routledge.

Fernández, J. (1988–89). Ocupaciones alfareras (2860 ± 160 años ap) en la Cueva de Cristóbal, puna de Jujuy, Argentina. *Relaciones N.S., XVII*(2), 139–178. (Buenos Aires: Sociedad Argentina de Antropología).

Fernández Distel, A. (1974). Excavaciones arqueológicas en las cuevas de Huachichocana, Dep. de Tumbaya, Prov. de Jujuy, Argentina. *Relaciones N.S., VIII,* 101–127. (Buenos Aires: Sociedad Argentina de Antropología).

Fernández Distel, A. (1980). Los fechados radiocarbónicos de la arqueología de la provincia de Jujuy. Fechas radiocarbónicas de la cueva CH III de Huachichocana, Tiuiyaco e Inca Cueva. *Argentina Radiocarbono en Arqueología I*(4/5), 89–100.

García, L. C. (1988/1989). Las ocupaciones cerámicas tempranas en cuevas y aleros en la Puna de Jujuy, Argentina–Inca Cueva, Alero 1. *Paleotenológica, 5,* 179–190. (Buenos Aires: C.A.E. A.).

García, L. C. (1998). *Arqueología de asentamientos formativos en la Puna Oriental y su borde, Provincia de Jujuy: El cambio hacia una vida crecientemente sedentaria y productiva en Azul Pampa, Departamento Humahuaca.* Ph.D. dissertation. Buenos Aires: Fac. de Filosofía y Letras, UBA.

Gepts, P. (2002). A comparison between crop domestication, classical plant breeding, and genetic engineering. *Crop Science, 42*, 1780–1790.

González, A. R. (1980). *Arte precolombino de la Argentina*. Buenos Aires: Filmediciones Valero.

González, A., & Pérez, J. (1971). *Primeras culturas argentinas*. Buenos Aires: Filmediciones Valero.

Goñi, R. (1995). El Uso Actual de Aleros: Algunas Implicancias Arqueológicas. *Cuadernos del Instituto Nacional de Antropología 16*, 329–341. Buenos Aires.

Hastorf, C. (1998). The cultural life of early domestic plant use. *Antiquity, 72*(278), 773–782.

Hastorf, C. (2003). Community with the ancestors: Ceremonies and social memory in the middle formative at Chiripa, Bolivia. *Journal of Anthropological Archaeology, 22*(4), 305–332.

Hayden, B. (1995). A new overview if domestication. In T. Price & A. Gebauer (Eds.), *Last hunters, first farmers* (pp. 273–299). Santa Fe: School of American Research.

Ingold, T. (2000). *The perception of the environment: Essays on livelihood, dwelling and skill*. London: Routledge.

Iriarte, J., & Vrydaghs, L. (Eds.). (2007). *Rethinking agriculture: Archaeological and ethnoarchaeological perspectives*. Walnut Creek, CA: Left Coast Press.

Jennings, J., & Bowser, B. (Eds). (2008). *Drink, power, and society in the Andes*. University Press of Florida.

Korstanje, M. A. (2005). *La organización del trabajo en torno a la producción de alimentos, en sociedades agropastoriles formativas (Pcia. de Catamarca, Rep. Argentina)*. Ph.D. dissertation. Tucumán: Facultad de Ciencias Naturales e IML, Universidad Nacional de Tucumán.

Korstanje, M. A. (2010). Producción y consumo agrícola en el Valle del Bolsón (1992–2005). In M. A. Korstanje & M. Quesada (Eds.), *Arqueología de la Agricultura: Casos de Estudio en la Región Andina Argentina* (pp. 48–75). Tucumán: Ediciones Magna.

Korstanje, M. A., & Quesada, M. (Eds.). (2010). *Arqueología de la agricultura: casos de estudio en la región andina argentina*. Tucumán: Ediciones Magna.

Korstanje, M. A., Quesada, M., Franco Salvi, V., Lema, V., Malobertti, M. (2015). ¿Cómo son los paisajes agrarios del formativo? In M. A. Korstanje, M. Lazzari, M. Basile, F. Bugliani, V. Lema, L. Pereyra Domingorena, & M. Quesada (Eds.), *Crónicas materiales precolombinas. Arqueología de los primeros poblados del Noroeste Argentino*, Sociedad Argentina de Antropología, http://www.saantropologia.com.ar/publicaciones/cronicas-materiales-precolombinas/

Korstanje, M. A., & Würschmidt, A. E. (1999). Producir y recolectar en los valles altos del NOA: Los Viscos como caso de estudio. In C. Aschero, M. A. Korstanje, & P. Vuoto (Eds.), *En los tres reinos: prácticas de recolección en el cono sur de América* (pp. 151–160). Tucumán: Instituto de Arqueología y Museo (UNT).

Latour, B. (2001). *La esperanza de Pandora*. Barcelona: Gedisa.

Lavallée, D., Julien, M., Karlin, C., Garcia, L., Pozzi-Escot, D., & Fontugne, M. (1997). Entre Desierto y Quebrada: Primeros resultados de las excavaciones realizadas en el abrigo de Tomayoc, (Puna de Jujuy, Argentina). *Bulletin del Institute Française de études andines, 26* (2), 141–175.

Lazzari, M. (2005). The texture of things: Objects, people and social spaces in pre-Columbian NW Argentina. In Lynn Meskell (Ed.), *Archaeologies of materiality* (pp. 126–161). Oxford: Blackwell.

Lema, V. (2008). ¿De qué hablamos cuando hablamos de domesticación vegetal en el NOA? Revisión de antiguas propuestas bajo nuevos abordajes teóricos. In S. Archila, M. Giovannetti, & V. Lema (Eds.), *Arqueobotánica y teoría arqueológica. Discusiones desde Suramérica* (pp. 97–126). Bogotá: UNIANDES-CESO.

Lema, V. (2009). *Domesticación vegetal y grados de dependencia ser humano-planta en el desarrollo cultural prehispánico del Noroeste Argentino*. Ph.D. dissertation. La Plata: Facultad de Ciencias Naturales y Museo, Universidad Nacional de La Plata.

Lema, V. (2010). Procesos de domesticación vegetal en el pasado prehispánico del noroeste argentino: estudio de las prácticas más allá de los orígenes. *Relaciones, XXXV*, 121–142. (Buenos Aires: Sociedad Argentina de Antropología).

Lema, V. (2011). Lo micro en lo macro: el tratamiento microscópico de macrorestos vegetales para la identificación de prácticas y modos de relación con el entorno vegetal en el estudio arqueológico de la domesticación vegetal. *Arqueología, 17*, 57–79. (Buenos Aires: Instituto de Arqueología, FFyL. UBA).

Madero, C. (1994). Ganadería incaica en el Noroeste argentino: análisis de la arqueofauna de dos poblados prehispánicos. *Relaciones N.S. XIX*, 145–163. Buenos Aires: Sociedad Argentina de Antropología.

Marx, K. (1946) [1867]. *El capital* (Vol. 1). México: Fondo de Cultura Económica.

McGuire, R. (1992). *A Marxist archaeology*. Academic Press.

Morrison, K. (1995). *Marx, Durkheim, Weber. Formations of modern social thought*. London: Sage Publications.

Núñez, L., & Dillehay, T. (1979). *Movilidad giratoria, armonía social y desarrollo en los Andes Meridionales: Patrones de tráfico e interacción económica (Essay)*. Antofagasta, Chile: Universidad del Norte. Dpto. de Arqueología.

Núñez Regueiro, V. (1975). Conceptos Instrumentales y Marco Teórico en relación al análisis del desarrollo cultural del NO argentino. *Revista, V*, 169–190. (Córdoba: Instituto de Antropología de Córdoba).

O'Shea, J. M. (1989). The role of wild resources in small-scale agricultural systems: Tales from the Lakes and the Plains. In P. Halstead & J. O'Shea (Eds.), *Bad year economics: cultural responses to risk and uncertainty*. Cambridge University Press.

Olivera, D. (1991). *Tecnología y Estrategias de Adaptación en el Formativo (Agro-alfarero Temprano) de la Puna Meridional Argentina. Un caso de estudio: Antofagasta de la Sierra (Provincia de Catamarca)*. Ph.D. dissertation. La Plata: Universidad Nacional de La Plata.

Oyama, S. (2002). The nurturing of natures. In A. Grunwald, M. Gutmann, & E. M. Neumann-Held (Eds.), *Human nature. Anthropological, biological and philosophical foundations* (pp. 163–170). Berlin: Springer.

Pérez Gollán, J. A., & Gordillo, I. (1994). Vilca/Uturuncu. Hacia una arqueología del uso de alucinógenos en las sociedades prehispánicas de los Andes del Sur. *Cuicuilco, 1*, 99–140.

Pochettino, M., Cortella, A., & Ruiz, M. (1999). Hallucinogenic snuff from Northwestern Argentina: Microscopical identification of *Anadenanthera columbrina* var. cebil (Fabaceae) in powdered archaeological material. *Economic Botany, 53*(2), 127–132.

Quiroga, L. (1999). "Los dueños de los montes, aguadas y algarrobales". Contradicción y conflictos coloniales en torno a los recursos silvestres. Un planteo del problema. In C. Aschero, M. Alejandra Korstanje, & P. Vuoto (Eds.), *En los tres reinos: prácticas de recolección en el cono sur de América* (pp. 217–226). Tucumán: Instituto de Arqueología y Museo (UNT).

Quiroga, L. (2010). En sus Huaycos y Quebradas: Formas Materiales de la Resistencia en las Tierras de Malfin. *Memoria Americana, 18*(2), pp. 185–209. (Buenos Aires: Instituto de Ciencias Antropológicas, Facultad de Filosofía y Letras, UBA/Sociedad Argentina de Antropología).

Raffino, R. A. (1973). Agricultura hidráulica y simbiosis económica demográfica en la Quebrada del Toro. Salta, Argentina. *Relaciones N.S., VII*, 297–332. (Buenos Aires: Sociedad Argentina de Antropología).

Rindos, D. (1984). *The origins of agriculture: An evolutionary perspective*. Academic Press.

Rodríguez, M. F. (1997). Sistemas de asentamiento y movilidad durante el Arcaico. Análisis de macrovestigios vegetales en sitios arqueológicos de la Puna Meridional Argentina. *Estudios Atacameños, 14*, 43–60.

Rodríguez, M. F. (1998). Resumen de la Tesis Doctoral: "Arqueobotánica de Quebrada Seca 3: Recursos vegetales utilizados por cazadores-recolectores durante el período Arcaico en la Puna Meridional Argentina". *Arqueología 8*, 161–168. Revista de la Sección Arqueología. Instituto de Ciencias Antropológicas. Facultad de Filosofía y Letras. UBA.

Rodríguez, M. F. (1999a). Movilidad e intercambios durante el Arcaico en la Puna Argentina. In C. Aschero, M. A Korstanje, & P. Vuoto (Eds.), *En los tres reinos: prácticas de recolección en el cono sur de América* (pp. 111–120). Tucumán: Instituto de Arqueología y Museo (UNT).

Rodríguez, M. F. (1999b). Plant species (Families: Poaceae, Asteraceae, Fabaceae and Solanaceae) at an archaeological site of the Southern Argentine Puna. *Journal of Ethnobiology, 19*(2), 228–247.

Rodríguez, M. F. (1999c). Arqueobotánica de Quebrada Seca 3 (Puna Meridional Argentina): Especies vegetales utilizadas en la confección de artefactos durante el Arcaico. *Relaciones, 24*, 159–184. Buenos Aires: Sociedad Argentina de Antropología.

Rodríguez, M. F. (2000). Woody plant species used during the Archaic period in the Southern Argentine Puna. Archaeobotany of Quebrada Seca 3. *Journal of Archaeological Science, 27*, 341–361.

Rodríguez, M. F. (2004). Cambios en el uso de los recursos vegetales durante el Holoceno Puna meridional argentina. *Chungará, 36*(1), 403–413.

Rodríguez, M. F., & Rúgolo de Agrasar, Z. (1999). *Deyeuxia eminens* (Poaceae: Agrostideae) en un sitio arqueológico de la puna meridional argentina (Provincia de Catamarca). *Darwiniana, 37*(3–4), 229–242.

Sayre, M. (2014). Ceremonial plants in the Andean region. In A. Chevalier, E. Marinova, & L. Pena-Chocarro (Eds.), *Crops and people: Choices and diversity through time*. European Science Foundation/Early Agricultural Remnants and Technical Heritage Program: Brussells.

Sponsel, L. E. (1989). Farming and foraging: A necessary complementarity in Amazonia? In S. Kent (Ed.), *Farmers as hunters. The implication of sedentism* (pp. 37–45). Cambridge University Press.

Tarragó, M. N. (1980). El proceso de agriculturización en el noroeste argentino, zona valliserrana. *Actas* del V Congreso Nacional de Arqueología Argentina (1978). Tomo I. San Juan: Fac. de Filosofía, Humanidades y Artes. UNSJ.

Vickers, W. T. (1989). Patterns of foraging and gardening in a semi-sedentary Amazonian community. In S. Kent (Ed.), *Farmers as hunters. The implication of sedentism* (pp. 46–57). Cambridge University Press.

Viveiros de Castro, E. (1996). Os pronomes cosmológicos e o perspectivismo ameríndio. *Mana. Estudos de antropologia social, 2*(2), 115–144. (Rio de Janeiro: Contracapa Ed).

Yacobaccio, H. (1990). Sistemas de Asentamiento de los Cazadores-recolectores Tempranos de los Andes Centro Sur. Ph.D. dissertation. *Buenos Aires. Facultad de Filosofía y Letras, Universidad de Buenos Aires*.

Yacobaccio, H., & Korstanje, M. A. (2008). Los procesos de domesticación vegetal y animal. Un aporte a la discusión Argentina en los últimos setenta años. *Relaciones, XXXII*, 191–215. (Buenos Aires: Sociedad Argentina de Antropología).

Zeder, M., Bradley, D., Emshwiller, E., & Smith, B. (Eds.). (2006). *Documenting domestication: Bringing together plants, animals, archaeology and genetics*. Berkeley: University of California Press.

Zimmerer, K. (1996). *Changing fortunes. Biodiversity and peasant livelihood in the Peruvian Andes*. Berkeley: University of California Press.

Chapter 5
Exploring Culinary Practices Through GIS Modeling at Joya de Cerén, El Salvador

Alan Farahani, Katherine L. Chiou, Rob Q. Cuthrell, Anna Harkey, Shanti Morell-Hart, Christine A. Hastorf and Payson D. Sheets

Introduction

Most archaeological sites are palimpsests of human activity, and interpreting the blurred, composite material traces of past daily practices can be challenging. The analysis of culinary practices in the past is equally complicated, as the organic residues and ceramic objects that are often utilized as evidence are not often found in situ at archaeological sites (Bray 2003; Fuller 2005). As a result, the spatial

A. Farahani (✉)
Cotsen Institute of Archaeology, University of California, Los Angeles, USA
e-mail: alanfarahani@ucla.edu

K.L. Chiou
Department of Anthropology, University of California, Berkeley, USA
e-mail: katie.chiou@berkeley.edu

R.Q. Cuthrell
Archaeological Research Facility, University of California, Berkeley, USA
e-mail: rcuthrell@berkeley.edu

A. Harkey
Department of Sociology and Anthropology, Wofford College, Spartanburg, USA
e-mail: harkeyae@wofford.edu

S. Morell-Hart
Department of Anthropology, McMaster University, Hamilton, Canada
e-mail: smorell@mcmaster.ca

C.A. Hastorf
Department of Anthropology and Archaeological Research Facility, University of California, Berkeley, USA
e-mail: hastorf@berkeley.edu

P.D. Sheets
Department of Anthropology, University of Colorado, Boulder, USA
e-mail: sheetsp@colorado.edu

M.P. Sayre and M.C. Bruno (eds.), *Social Perspectives on Ancient Lives from Paleoethnobotanical Data*, DOI 10.1007/978-3-319-52849-6_5

relationships between preserved culinary objects rarely lend themselves to fine-grained approaches that consider the culinary dimensions of lifeways at the level of lived experience. Nevertheless, just such an opportunity is afforded by the archaeological site of Joya de Cerén, thanks to its level of preservation and unique circumstances of site formation.

Joya de Cerén, sometimes referred to as the "Pompeii of the New World," was discovered in southern El Salvador in 1976 when a bulldozer exposed part of a domestic structure. Since 1978, three decades of archaeological work (Sheets 2002a) have established that the site was buried under nearly five meters of volcanic tephra due to the eruption of the Loma Caldera volcano (Miller 2002) on an August evening between 610–671 C.E. (McKee 2002). The Cerén community likely fled prior to its destruction, as human remains have not been found at the settlement. It does not seem, however, that the community had enough time to take their possessions with them—most objects were left in the exact places they were used (McKee 1999). These objects include ceramics, worked bone, ground stone tools, chipped stone tools, and figurines, often found within collapsed and yet well-preserved wattle-and-daub structures. The volcanic eruption also preserved a great deal of the plants consumed by the local community, both the domesticated plants grown in and around the site, as well as the wild varieties (Lentz et al. 1996; Lentz and Ramirez-Sosa 2002).

Through the instant preservation of the materials of everyday life discarded exactly in their location of use, Cerén has provided a research team led by Payson Sheets the rare opportunity to disentangle the ambiguous patterns that typify archaeological datasets in order to answer questions directly related to the daily lifeways of Cerén's past inhabitants (Sheets 2000, 2002a). Because of these unique conditions, it presents an ideal case in which to model the spatial dimensions of past culinary lifeways. Since the spatial relationships between culinary objects at Cerén almost entirely correlate to their original use-contexts, it is possible to employ spatial database-based approaches, such as that provided by GIS, to query associations of culinary objects at the scale of community and household. Geographic Information Systems (GIS) software has been widely applied in archaeology, especially in regard to landscape studies (Siart et al. 2008), least-cost pathways (Taliaferro et al. 2010), and predictive settlement modeling (Church et al. 2000). Yet the potential of manipulating spatially organized datasets for intra-site exploratory data analysis has been less-often realized (for notable exceptions, see Boudreaux 2007; Van Derwarker et al. 2014; Wilson 2008). As a result, one of the more basic scalar units of archaeological research—the excavated site—still has the potential for continued methodological development vis-à-vis spatial database modeling (Neubauer 2004).

In collaboration with the project director Payson Sheets and the original excavators and specialists at Céren, a research team at UC Berkeley under the direction of Christine Hastorf created a spatial database based on the site's rich archaeological dataset. The primary motivation was to explore the aforementioned relationships between daily culinary practice, household economies, and spatial structuration. The construction of a spatial database, a secondary objective of the project, enabled analyses and visualizations of object associations impossible through other means. The Céren project therefore has, and continues to have, multiple research objectives, only one of which is discussed below.

Research Design

The research design for this iteration of the Cerén GIS Project was predicated on investigations by the original research team (under Payson Sheets) of household economies related to culinary practices and the settlement-wide distribution of task-areas in particular. This orientation is drawn from the term *taskscape*, a concept developed by Ingold (2000) that views the spaces within which activities are performed not as static, strictly bounded locales, but rather as fluid and mutable areas, analogous to understandings of landscape—for "just as landscape is an array of related features, so—by analogy—taskscape is an array of related activities" (1993: 158). Taskscapes are "the entire ensemble of tasks, in their mutual interlocking" (2000: 158), understood as having spatial aspects for the purposes of analysis, a conceptual notion that archaeologists are capable of utilizing to great effect. Using the concept of *taskscape*, focus is placed on the practical operations of households at Cerén, the "constitutive acts of dwelling" (Ingold 2000: 158) that take the form of an array of activities over time. Since archaeologists excavate dwellings and domestic objects rather than social groups, the excavators inferred households from the relative clustering or dispersal of structures and features at Cerén (Wilk and Rathje 1982). Households were numbered according to order of excavation, and the associated objects and features were assigned membership accordingly (Sheets 2002a). Although the site appears to be a static snapshot in time of the types of activities people were engaged in during the moments before the volcanic eruption, underlying and informing these activities were the practices, materials, and historicity of a dynamic taskscape.

During the course of excavation, site excavators recorded the presence of production-related objects in an area designated Household 1. Examples include grinding platforms (*metates*), grinding stones (*manos*), ceramic vessels, and the seed remains of food plants (Sheets and Simmons 2002: 181). The metates were likely utilized in food preparation, specifically the grinding of seeds such as maize kernels (Beaudry-Corbett et al. 2002: 56). The GIS project explored the associations of objects such as metates with other objects in their immediate vicinity, hypothesizing that locations with metates were active areas of food processing for individuals within households, and potentially primary foci for food distribution. The co-occurrence of other objects or organic remains can be linked with food production, distribution, and storage activities—all argued to be particularly important in Household 1 (Sheets and Simmons 2002: 181).

Establishing the Cerén GIS Spatial Database

Due to the impracticability of analyzing the site's entire extent, which includes several other households, only the Household designated as "1" was sampled in this study. Household 1 represents the most completely excavated set of structures at Cerén, with five structures containing carefully provenienced objects, faunal, and botanical remains (Beaudry-Corbett et al. 2002). The Cerén GIS began with several objectives in mind: first, to produce an accurate, rectified cartographic representation of the excavated area, including accurate data-as-points, and second, to use the subsequent spatial database to carry out exploratory data analysis and employ spatial statistics (Mayer 2006). The first goal was executed in three stages: (1) mapping all structures, (2) delineating distinct architectural and environmental features, and (3) plotting all objects and organic remains with their attribute data. This created the spatial database necessary to apply inferential and descriptive spatial statistics. The second goal explored this spatial data (Andrienko and Andrienko 2006) through density analyses of certain object groups as well as with proximity-based measures on the positions of objects and organic remains in this area.

To establish a cartographic representation of the settlement, the team began with published data and unpublished field reports by the excavators and specialists of Cerén (Sheets 2002a; Sheets and Brown 1996; Sheets and Kievet 1992; Sheets and McKee 1989, 1990; Sheets and Simmons 1993). With their assistance, the team assembled spreadsheets containing attribute data for three object classes (ceramic, groundstone, and chipped stone), two classes of organic remains (botanical and faunal), site drawings at multiple scales of resolution, and GoogleEarth imagery. ESRI ArcGIS 10.1 software was utilized as the spatial database and as the cartographic platform for spatial analyses and site visualizations. A multistep procedure was employed to incorporate the multiple complementary datasets; the first step involved transforming the published map of Cerén into a working tool interpretable by GIS. Due to the presence of a protective hangar over the site, absolute projected coordinates for the site datum could not be established. Instead, an arbitrary projection was created in UTM using a local site datum. This provided the anchor for the base map (based on Sheets 2002a: 2) upon which the structure maps produced by excavators were overlain. Each structure map (found in the unpublished *informes*, or excavation reports) was anchored to the base map using the excavators' original grid system, when available. For structures that were mapped before the grid system was established, anchor points were judgmentally attached onto features on the site-wide base map. It is anticipated that the representation of certain structures may not precisely reflect their real-world positions, but through adjustments to the structure positions the error margin (calculated through georeferencing in GIS) lies within a few centimeters.

Within the current Cerén GIS framework (Fig. 5.1), structure walls are represented through polygons, and archaeological contexts around architectural features and structure areas are layered, operationalized as "floors", "patios", and "benches",

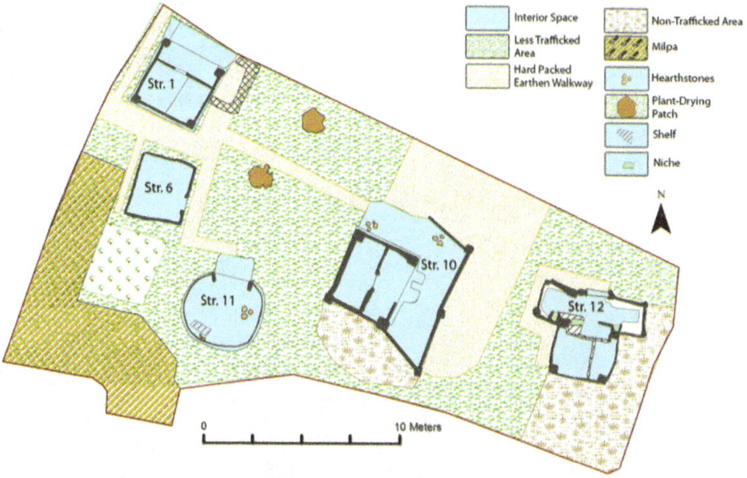

Fig. 5.1 A GIS representation of Household 1 at Cerén with a legend providing details regarding the illustrated archaeological contexts and features. The *red line* around the Household indicates the extent of excavation. The hatched addition shown on Structure 1 represents a low wall, and the green polygon to the right of the milpa (maize field) on the west side of the Household is a kitchen garden. A one meter drip-line is not shown, but surrounded each structure, except for Structure 12, which contained a 30 cm drip-line

and defined within the database as "exterior" or "interior" spatial locations. Surrounding features such as gardens, plant drying patches, and hearths were also incorporated and represented as polygon features. A 1.5 m buffer was calculated for each structure to simulate the extent of the thatched roofing (see Fig. 5.2), given the extent of the drip lines encountered during excavation (McKee 2002: 60) and the extent noted historically (e.g., Wauchope 1938). The objects and organic remains are represented as point data, as each item possesses precise and discrete X, Y coordinates. The attribute data for each of these classes was appended to each object. In many cases, the attribute data was derived from multiple sources. For instance, the paleoethnobotanical data was derived from publicly available Field Specimen Lists (http://Cerén.colorado.edu/Cerén_FS_Lists/readme.html), published reports (Lentz et al. 1996; Lentz and Ramirez-Sosa 2002), and collaboration with excavators and analysts (Lentz and Sheets, *pers. comm.*). Each object therefore contained different attribute data specific to its class. With all of these data sources assembled (and always in the process of improvement through additional clarification and data), a query-able spatial Cerén database was established.

Initial investigations of Cerén using this GIS database employed spatial statistics to explore possible "task-spaces" in Household 1 on a general, exploratory level. These more exploratory analyses focused on locating the spatial patterning of select object classes using average nearest neighbor analyses as well as the clustering of object-specific values using the Getis-Ord Gi* statistic (Getis and Ord 1992).

Fig. 5.2 General distribution of ceramics in the structures of Household 1. *Dark red* and *dark green* color contour lines indicate areas with >2.0 vessels/m^2, based on a kernel density analysis of only ceramic vessels identified as bowls or jars within a 1 m radius. *Light orange* stippling outside of structure walls indicates schematized structure drip lines

Tentative results included the clustering of botanical remains found in ceramic vessels in Structures 1 and 6. Likewise, most of the clustering of objects and organic remains seem to occur within structures, rather than without, which points to structures as the center of activities, especially during the hot, humid temperatures brought on in August during the mid-rainy season (cf. Sheets and Woodward 2002: 189).

The analyses performed for this study were less focused on the application of spatial statistical techniques on the existing dataset, and more focused on modes of dynamically querying the available data to understand the relationship between built spaces and household economies and culinary practices. Although similar analyses have been carried out elsewhere in Mesoamerica (e.g., Ashmore and Wilk 1988; Fletcher 1983; Hendon 1996; Levi 1996; Manzanilla and Barba 1990), the richness and condition of the Cerén dataset allows for more confidence in potential inferences usually impossible at sites with less remarkable preservation.

Culinary Practices: Bowls, Jars, and Small Household Tasks

One way to analyze household spaces is to contrast areas that were used for active tasks, such as food preparation or consumption, with areas used mainly for storage. This allows consideration of both in what specific locations certain activities were taking place, and also how the presence, division, or overlap of different activities can illuminate the nature of each space, and the types of functions different structures had. In *Before the Volcano Erupted: The Ancient Cerén Village in Central America*, Beaudry-Corbett et al. (2002) addressed this topic by discussing the distribution of jars and bowls across households. She suggested that bowls (smaller, lower, and more open vessels) were associated with food processing, consumption, and very small-scale transport. Meanwhile jars (larger, taller, more closed vessels) would be best suited to stationary storage of large quantities of material, or, in the case of jars with handles (or those of a smaller size and weight) well suited to both stationary storage and transport of larger quantities over longer distances (Beaudry-Corbett and Bishop 2002: 122). Thus in an area reserved primarily for storage, we would expect to see large, storage-type jars almost exclusively. In an area used for food consumption (but not for storage or processing) we would expect mainly bowls, and perhaps an occasional small jar from which food would be served. In an area used for processing, there might well be an overlap of bowls and jars of various sizes. Likewise we might expect overlapping types in places where many different activities took place.

A preliminary density raster analysis of ceramic bowls in Household 1 shows that within most of these structures in this household, bowls are either found singly or in groups of two to three. In this analysis, we used kernel density rather than point density, as it uses a kernel function to create a smooth surface between points to calculate relative density (Fotheringham et al. 2000: 45–49). In the absence of any known patterning to the data, a search radius of 1 m (cell size = 0.05 m) was chosen. This value was meant to approximate a minimum search space within a "casual" grabbing and leaning distance around the remains.

Within Household 1, groups of bowls tend to be closer to walls rather than in the centers of rooms. These bowls were sometimes nested, suggesting they were not in active use. A separate density analysis of jars (represented in red) shows a slightly different pattern than bowls, usually clustering in much larger groups. Three of the Household 1 structures (1, 10, and 11) have groupings of seven or more jars, always along a structure wall. The storehouse Structure 6 does not follow this pattern, however, with a cluster of jars in the center of the room. This could reinforce the interpretation of the space as not in use at the time of the eruption as a locus of daily activities.

By overlaying bowl and jar density, it can be seen that ceramics tend to be segregated and clustered by type in the domicile Structure 1 and the "ritual" Structure 12, but mixed in the other Household 1 structures (Fig. 5.2). This pattern suggests that in some areas, food was stored and consumed but probably not actively processed. When examining the specific loci of activities, we also see that the Structure 1 domicile and the Structure 12 ritualized location were marked by more highly differentiated activities, rather than frequently overlapping practices seen in other structures. To explore the complex patterns in practices represented by bowls and jars in the other structures, other object types and attributes were considered.

Culinary Sets: Clustering of Materials

In addition to the investigation of the distribution of jars and bowls in the context of food preparation and storage in active spaces, characteristics of vessels with close provenience were also explored to identify culinary practices across the structures of Household 1. To address this topic, the aforementioned spatial density analysis was used to delimit "culinary sets" based on especially dense areas of jars and bowls (seen above). Cluster analysis was used to compare these "culinary set" contents, and associations were then examined between vessels and their botanical contents.

One attribute recorded for most ceramics was the presence or absence of handles. Following Redfield (1950), we proposed that jars and bowls with handles were easier to transport and were probably more actively used in tasks involving heavier materials, large quantities of materials, or materials that were transported longer distances. To explore sets of ceramics used together in culinary practice, a second kernel density raster (radius = 0.5 m; cell size = 0.05 m) was generated of all ceramic objects with secure provenience information in Household 1 (n = 119; Fig. 5.3). Associated culinary sets were defined as all ceramic objects within areas that met the following conditions: (a) kernel density in the area was greater than 3.0 vessels/m^2; (b) the area contained greater than four vessels; (c) the area was entirely in interior or exterior space (i.e., walls separate sets); (d) areas could not be greater than ca. 2.5 m in diameter. Vessels not identified as either jars or bowls were included in the density raster, but are excluded from interpretation. Table 5.1 summarizes eight culinary sets containing a total of 59 vessels that were identified in Household 1. The locations of culinary sets included in this analysis are indicated by dashed lines in Fig. 5.3, and the contents of each culinary set are presented in Table 5.1. To visually summarize relationships between culinary sets, a hierarchical cluster analysis (Ward's method, Ward 1963) was performed using counts of the following four categories: jars with handles, jars without handles, bowls with handles, and bowls without handles (Fig. 5.3, top right).

Structures 11 and 6, interpreted as a kitchen and storeroom, respectively, each contained three culinary sets. Both structures contained two culinary sets that were grouped together in Cluster 3 of the hierarchical cluster analysis (sets 6A, 6B, 11A, and 11B) and one culinary set in Cluster 1 (sets 6C and 11C; see Fig. 5.3). In both structures, the two culinary sets from Cluster 3 were adjacent to one another. Culinary sets in Cluster 3 contained exclusively jars with handles (or no jars) and

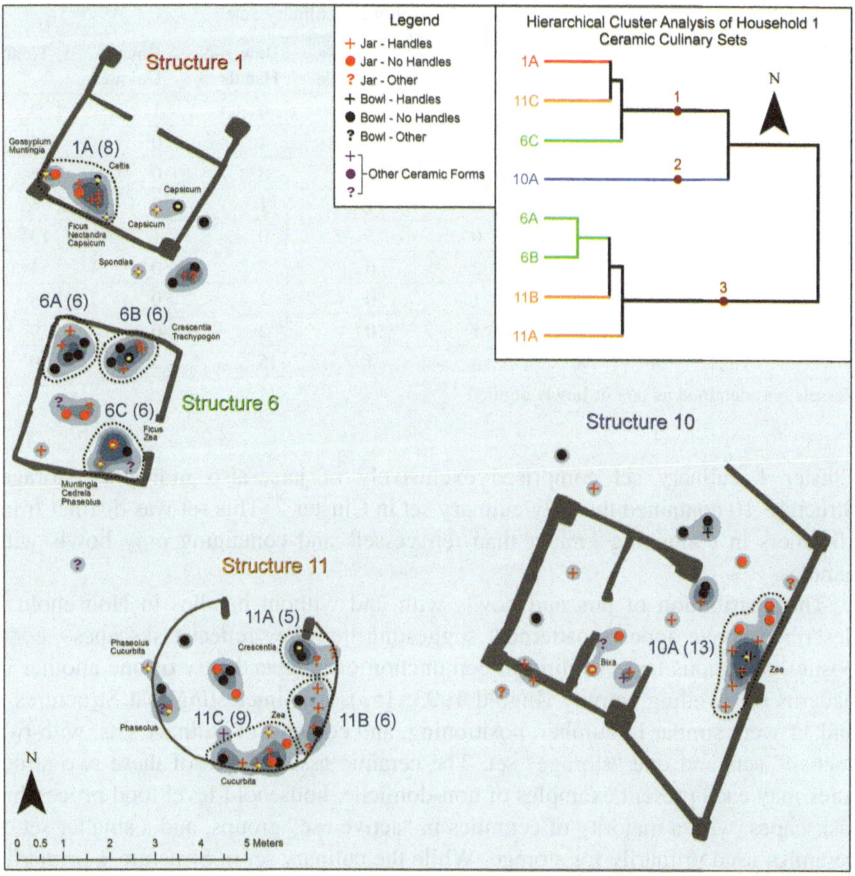

Fig. 5.3 Distribution of ceramics in Household 1 structures. No culinary sets were identified in Structure 12 (not shown). *Dark blue* color contour lines indicate areas with >3.0 vessels/m^2, based on a kernel density analysis of all ceramic vessels with 0.5 m radius. Vessels are color-coded according to form (*red* = jar; *black* = bowl; *green* = unknown form) and assigned a symbol according to presence or absence of handles (*cross* = handles present; *circle* = handles absent; *question mark* = presence or absence of handles unknown). Culinary sets are indicated by *dashed lines* and identified in *dark blue* lettering with number of vessels in parentheses. *Top right*—hierarchical cluster analysis (Ward method on a Manhattan distance; dendrogram scale proportional to actual cluster distance) of contents of culinary sets based on the following four categories: jars with handles, jars without handles, bowls with handles, and bowls without handles. Samples included in cluster analysis are color-coded by structure number. Numbered *red dots* on the cluster analysis dendrogram identify clusters used in interpretation

bowls without handles, and bowls were generally more abundant than jars (Table 5.1). Sets in Cluster 1 were more diverse, with greater numbers of jars than bowls and jars both with and without handles. Culinary sets in Cluster 3 may represent areas of more active processing, while those in Cluster 1 may have been for longer term storage. Structure 1, interpreted as a domicile, contained one

Table 5.1 Categories of ceramic vessels in Household 1 culinary sets

Structure	Cluster	Jar—Handle	Jar—No Handle	Jar—Unknown	Bowl—Handle	Bowl—No Handle	Bowl—Unknown	Total
1	1A	4	2	1	0	0	1	8
6	6A	2	0	0	0	4	0	6
6	6B	2	0	1	0	3	0	6
6	6C	3	1	0	1	1	0	6
10	10A	6	3	0	3	0	1	13
11	11A	0	0	2	0	3	0	5
11	11B	3	0	1	0	2	0	6
11	11C	4	2	1	0	2	0	9
	Total	24	8	6	4	15	2	59

Vessels not identified as jars or bowls omitted

Cluster 1 culinary set comprised exclusively of jars, also indicating storage. Structure 10 contained the only culinary set in Cluster 2. This set was distinct from all others in containing greater than ten vessels and containing only bowls with handles.

The distribution of jars and bowls with and without handles in Household 1 described above appears patterned, suggesting that they reflect taskscapes—composites of various tasks cycling in conjunction or independently of one another in patterns of dwelling activity (Ingold 1993: 153). It is interesting that Structures 6 and 11 were similar in number, positioning, and contents of culinary sets, with two "active" sets and one "storage" set. The ceramic assemblages of these two structures may each present examples of non-domicile, household-level food processing taskscapes, with a majority of ceramics in "active use" groups, and a smaller set of ceramics used primarily for storage. While the culinary set in Structure 1 primarily indicates storage, that of Structure 10 may represent a different kind of taskscape associated with inter-household food processing and consumption, as indicated by the larger numbers of overall vessels, and particularly of more easily transportable serving vessels (bowls) with handles. This accords with Sheets' (2002a) interpretation of Structure 10 as a communal structure associated with food consumption.

Culinary Stations: Tasks Near Metates

Since the food processing and subsequent storage of processed remains often occurs in specialized taskscapes, the next analysis was to utilize the locations of stationary metates as valuable heuristic analytical units. Metates are grinding stones of various sizes and elaboration, found throughout Mesoamerica in elevated contexts (such as low tables), directly atop ground surfaces, discarded in refuse areas, or incorporated into architecture (e.g., Plunket and Uruñuela 2000). Grinding stones may be used inside or outside the home (Clark 1988). Metates, according to Michael Searcy, are "costly tools, valued by family members over many generations" (2011: 3), sometimes passed on to newlyweds (2011: 72), and kept for up to 150 years (2011: 73; see also Webster et al. 1997: 57). They may be used periodically over the year or multiple times throughout the day (Searcy 2011: 76).

Grinding stones can reveal information including which foods were processed, how foods were processed, intensive or infrequent use of locations, local or distant sourcing of materials, and shifts in style over time (Clark 1988; Hayden 1987; Schneider 2002: 92). Answering such questions can build into discussions of lifeways, economic networks, and social patterns, as well as nutritional status, group identity, and enculturation (Moholy-Nagy 2003; Schneider 2002: 10). Object biographies of metates can address "the material that was chosen for the tool and why; the design of the object; how the object was used (and possibly recycled); [and] why and how the object was discarded" (Schneider 2002: 92; similar to Searcy 2011: 4). In archaeological contexts, it is difficult to tell the difference between a metate left in-use, resting, stored, or discarded as they are usually heavy and difficult to transport (following Plunket and Uruñuela 2000: 81) and thus may remain stationary regardless of status at time of abandonment.

Like the majority of objects at the settlement, metates at Cerén are most often found in elevated contexts, as opposed to resting in contact with the floor (Sheets 1998: 66). None of the metates recovered in Household 1 are in discard areas or incorporated into architecture, so they may be considered provisionally in-use, that is, "belonging to the world of the living" (following Plunket and Uruñuela 2000: 81). In earlier studies, Sheets has associated metates with individuals, households, and broad economies, while Sweely (1998) has addressed power relations implicated by locations of activities including maize grinding. Given that "the predominant pattern in traditional Maya villages is for an economically active female to have a single *mano* and *metate*" (Sheets 1998: 74), Sheets has interpreted the lone mano-metate set in Household 4 as likely belonging to "one economically active female" (Sheets 1998: 70). He has compared this situation with that of four active women in Household 1, where several metates were mounted on *horquetas* (forked sticks) while another was in contact with the floor (Sheets 2000: 225). One metate is a miniature, palm-sized, and legged form, which Sheets (2000: 221) suggests was used for grinding hematite, perhaps for use as body paint.

Of the four metates likely used for maize, Sheets hypothesizes that three, demonstrating light use-wear (Sheets 2002b: 148), were used only as "overflow" for large-scale events related to the nearby Structure 10 (Sheets 2000: 225; 2002b: 148), activities noted ethnographically as related to grinding mills in modern town centers (Searcy 2011: 30). Sheets implicates extra-household economies, believing that the corn undergoing grinding in the Household 1 area, from the three metates exhibiting light use-wear, was "perhaps to feed participants in religious rituals" (1998: 74). If this is the case, surrounding materials may also correspond with this "episodic household craft specialization in service to a religious organization" (Sheets 1998: 74). (Another hypothesis is that that new grinding stones were acquired and stored in anticipation of passing along the "heirloom" implements to children when marrying, a modern practice documented by Searcy 2011: 138.) The fourth *metate* used for maize grinding was found resting on the ground of the kitchen structure (Structure 11), adjacent to a three-stone hearth and a vessel containing soaked maize kernels (see Fig. 5.1).

Thus, as heavy- and less-portable fixtures in the structures of Household 1, metates index fairly stable culinary stations, active spaces where the people of Cerén ground maize, seeds, and other organic and inorganic goods. When the distribution of vessels and botanical remains are visualized together, it is possible to see that metates tend to be located within the clusters of jars with handles (see, *inter alia*, Fig. 5.4b). In essence, the people of Cerén who were engaged in grinding activities appear to have kept easily movable jars containing soon-to-be processed foods within arm's reach. In contrast, jars without handles are located at slightly greater distances from these "grinding stations" with only one exception. These more distant jars contain different types and smaller quantities of materials. Of these vessels, Sheets has posited that "some may have been used to transport the soaked maize kernels for grinding, others may have been used to catch the ground *masa*, and the polychrome bowl may have been used as a food serving vessel" (Sheets 1998: 85).

To further explore nodes of activities related to metates, two near analyses were performed, initially using a 0.5 m small-scale radius as a heuristic bootstrap. This 50 cm radius defines a hypothetical arm-length distance from a metate while seated. Near Analysis calculates the distances between some point feature and any number of other features. Using the output of the Near Analysis Table, it was possible to generate a map of the locations of the objects that occur within 50 cm of the metates (Fig. 5.4a, b). Unsurprisingly, the closest items to two of the three metates on the western portion of Household 1 were fragments of manos. Likewise, in the kitchen structure (Structure 11), the three objects closest to a metate near the kitchen hearth were a complete mano, a ceramic jar with handle, and a bowl (Fig. 5.4b).

The association of these objects in such a spatially circumscribed zone implicates metates as locales of not only food production but also of food distribution. The propinquity of a ceramic jar with handles, again the most likely to contain seed remains, is probably not accidental. More illustrative is how relatively free of objects the spaces around metates seem to be. The metate in the storehouse, Structure 6, is near another metate fragment and three bowls. The "neatness" of this space may be a feature of the Cerén taskscape at this particular moment in time. Sheets has posited that, along with areas at greater distances from structures (at least a few meters), metate preparation areas were "single-function activity areas" (1998: 94), whereas most other areas at Cerén were multifunctional, with clean-up after each activity (1998: 94).

Given that 50 cm is an arbitrary search radius, the near analysis was extended by another 50 cm in order to examine the differences in the kinds and numbers of objects that appear in between these spaces (Fig. 5.4). With few exceptions, the enlargement of the search space only yielded more ceramic vessels, the majority of which were handle-less bowls, but also including more jars with handles. Considering that these items are within "leaning" distance of a seated individual,

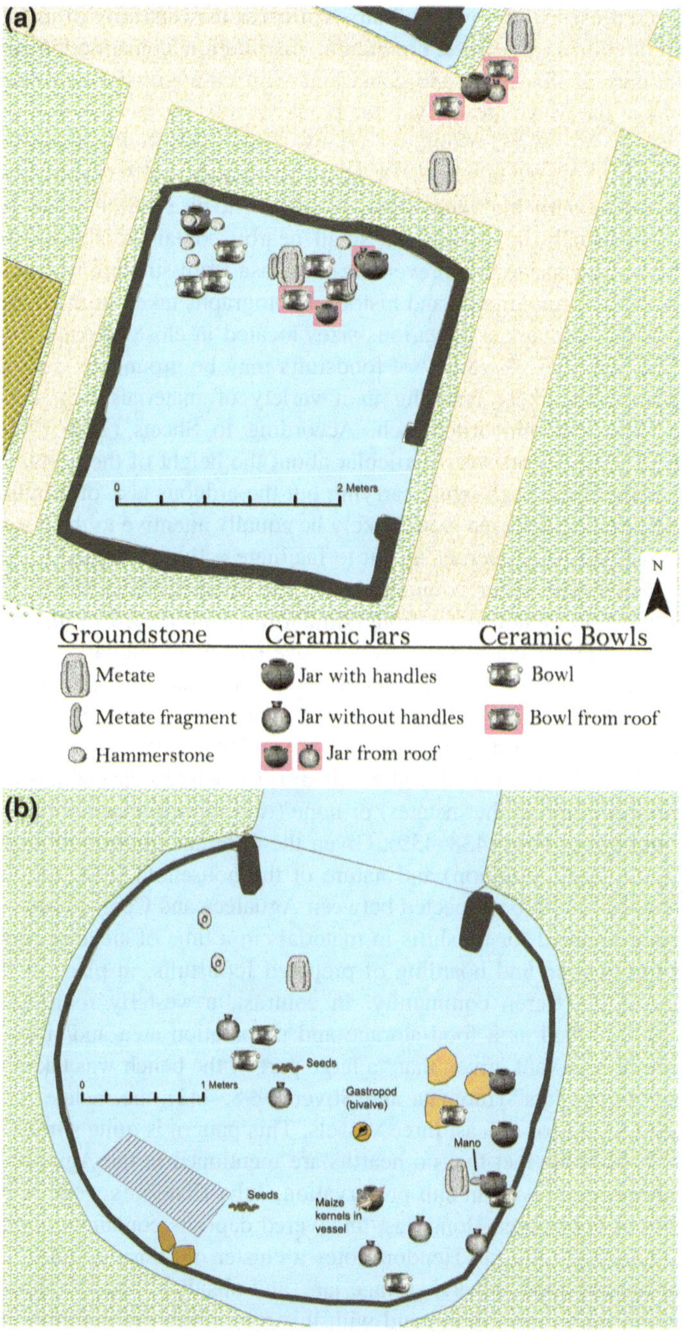

Fig. 5.4 Distribution of objects within a 1 m radius in select structures in Household 1. *Panel a* represents Structure 6, and *panel b* represents Structure 11. The legend for the symbology of the recovered objects is in between the two images, and symbols not represented in the legend are supplemented by on-graphic text

the additional presence of bowls and jars reinforces the centrality of metates within a spectrum of culinary practices: production, distribution, consumption and storage.

The findings of this paper are consistent with ethnographic and ethnohistoric descriptions in the Maya area of various practices relating metates to food storage, preparation, serving, and consumption (Hanks 1990: 331; Redfield and Villa Rojas 1934), as well as finger bowls for washing [de Landa 1978 (1566): 34]. Searcy notes that for modern highland Maya women who use metates, the process of grinding corn usually incorporates a small nearby container of water for use in wetting the soaked maize and preventing the masa from sticking to hands (Searcy 2011: 114, 119). Ethnographic and historic photographs taken in the region almost inevitably depict containers of various sizes located in close proximity to metates and grinding activities. A variety of foodstuffs may be ground on a single metate (Searcy 2011: 76), likely resulting in a variety of materials kept conveniently nearby in vessels within arms-reach. According to Sheets (1998: 74), modern women in El Salvador are very particular about the height of the grinding surface, given the angle of the back while carrying out the arduous task of grinding. Those involved in grinding at Cerén would likely be equally attentive to the positioning of the metate and various materials meant to facilitate activities related to metate use.

As compared with other communities in the ancient Maya world, the site of Cerén continues to demonstrate strong similarities with regard to metates as culinary stations. Inomata and Stiver (1998) recovered sets of culinary materials at the rapidly abandoned site of Aguateca, Guatemala. The metate in the elite household associated with structure M8-10 was located near large ceramic jars "used for the storage of food and liquid" (Inomata and Stiver 1998: 438). Figures in their text reveal that metates were located well within arm's reach (50 cm) of only one or two ceramic vessels (three of the metates) or none (two of the metates) in this structure (Inomata and Stiver 1998: 438–439). Given the different contexts of abandonment (warfare vs. volcanic eruption) and nature of the household (elite vs. commoner) some differences are to be expected between Aguateca and Cerén. These variations are perhaps accounted for by shifts in materials in a time of siege at Aguateca for structural fortification and hoarding of prepared foodstuffs, in place of the immediate flight of the Cerén community. In contrast, a westerly room of structure M7-35 was described as a food storage and preparation area and/or possibly the living space of a servant, given that "a large part of the bench was taken up by the metate and storage jars" (Inomata and Stiver 1998: 442), the metate lying within arm's reach (50 cm) of at least three vessels. This pattern is quite similar to that of Cerén, aside from the fact that no hearths are mentioned in the Aguateca study.

In another context with in situ preservation, Julia Hendon's work at Patio B of Group 9N-8 at Sepulturas, Honduras, uncovered deposits containing entire objects left under building collapse. Hendon notes a cluster of "three metates, a portable brazier for heating food, several storage jars, and obsidian blades" (Hendon 1997: 36). Although no figures correspond with this text, given the narrative it is likely that the metates were stationed within arm's length of the nearby objects. Less pristine contexts have also demonstrated the clustering of vessels with metates. At the site of Piedras Negras, in Guatemala, for instance, a nearly complete metate

(PN-46F-8) was found in one of the rooms of structure J-33, near a broken but complete vessel (Golden 2002: 251–252). Overall, however, it remains difficult to recover additional evidence of metates in situ, as ethnographic evidence documents either the sale or removal of valuable groundstone items prior to departure from a residence (Lange and Rydberg 1972). Moreover, several works have documented the retrieval and reutilization of pre-Columbian metates by modern populations (e.g., Lange and Rydberg 1972: 430; Hartman 1907: 39 cited in Lange and Rydberg 1972: 430–431).

In imagery depicting grinding stones, the pattern of vessels located near metates continues. It is rare, however, to find depictions of food preparation in Classic period art (Houston et al. 2006: 107), much less the specific task of grinding on a metate. At least one plate depicts a vessel located just next to a metate [JM03204 in Montgomery (2000)]. Several ceramic dishes depict metate use by women, in each case with at least one nearby vessel (Kerr 631 in Houston et al. 2006: 111; Kerr 1272 in Coe and Kerr 1982: 94). Two-dimensional representations, however, appear to differ from three-dimensional representations. The few examples of sculpture and figurines depicting the use of metates do not tend to include nearby vessels, but are usually comprised simply of a woman grinding with a metate and mano. In Central Mexico, the Borgia, Florentine, and Mendoza codices all depict maize grinding on a metate with one or more vessels within easy reach. The Codex Mendoza helpfully labels the vessels near a metate, in a panel depicting children completing various chores (Berdan and Anawalt 1992: folio 59V). In this panel, a young woman is seated next to a three-stone hearth with a *comal* atop it, while an *olla* and a tripod *escudilla* vessel sit within easy reach.

Given the evidence presented by the nearest neighbor and density analyses in GIS, it is apparent that certain groundstone tools and vessels are strongly associated with metates. When compared with other communities and time periods, the combined results build a strong case for metates as persistent culinary stations; taskscapes that incorporated a variety of materials and implements, conveniently kept within close range.

Culinary Agents: What the Duck Saw

In addition to the investigation of household-wide spatial activities, it is possible to explore the relationship of objects and organic remains on the scale of an individual structure, and more finely, even within the spaces of a structure. One particularly poignant example of this can be seen in the corner of Structure 6: while no human witnesses remained in Cerén at the time of the eruption, a single domestic duck was left behind, tied by one foot to a pole in the wall of the structure.

Just as spatial analyses of primary contexts have offered glimpses into activities of the people who built these homes, even the patterns of a duck's experience are

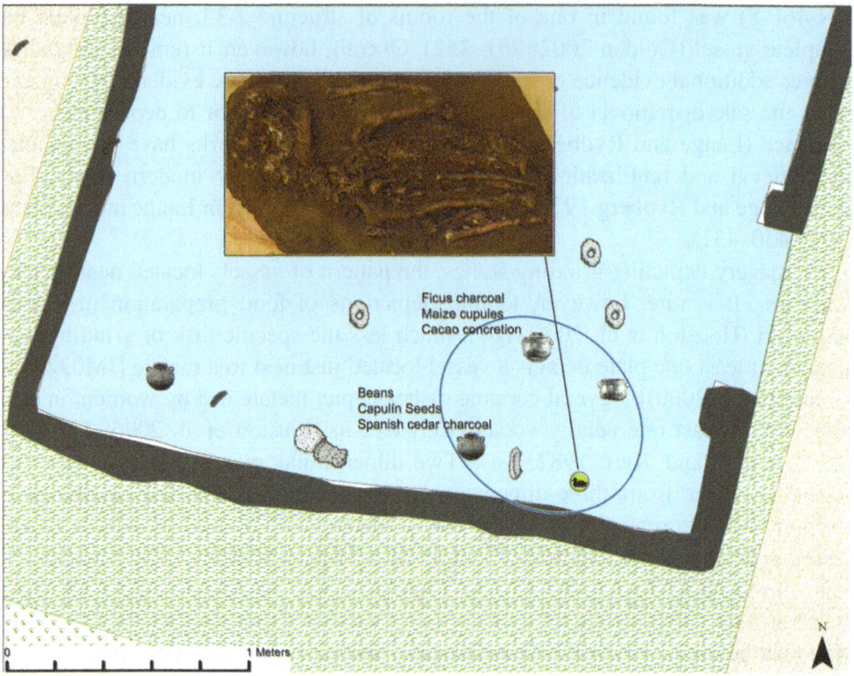

Fig. 5.5 Distribution of objects and organic remains (labels represent anatomical parts of a plant or general type—e.g., charcoal) around remains of a duck (Anatidae) which is symbolized by a *yellow circle* with a black duck silhouette. The *blue circle* represents the object "cluster" discussed in the text. The other symbols correspond to those in Fig. 5.4

visible (Fig. 5.5). This corner of Structure 6 is largely cleared of clutter as compared to other areas in the same structure, but the few nearest objects suggest an animal carefully tended. Nearest to the duck is a bowl of beans braced by a metate fragment and set at a comfortable height for the animal. The bracing at the base perhaps lent extra stability so that the duck would not tip the bowl. A second small bowl was empty at the time of excavation, but may well have been the vessel for the duck's water. Slightly farther away, though still in reach, is a stack of three bowls with a quantity of corn at similar height. Was the duck so well looked-after as to have multiple food sources at its disposal?

The bird was kept only a few feet from what seems to be an active work area, or the storage space of active culinary equipment, with mano and metate and numerous bowls and jars. One wonders if someone working in that space regarded the duck as a pet, a bit of company during tedious jobs. Perhaps someone from the Cerén community may have felt a twinge of sadness at leaving the bird to its fate. Or perhaps these foods were simply meant to rapidly fatten a duck intended for the stewpot, and tragedy lay only in the fact that delicious poultry went to waste.

Conclusions

The combination of density analyses and near analyses provides strong evidence that food production, storage, and distributive activities at Cerén were organized into spaces that were relatively discrete and clustered. The distribution of jars and bowls points to a nonrandom distribution of these objects that played a role in how the Cerén community organized their productive space. Jars and bowls occupied clearly divided spaces, such as in Structure 1, where their active separation might have been related to their intended use or even ideas about how, and where, such materials must be temporarily stored. Metates seem to be in close proximity with jars with handles, a ceramic type spread throughout Household 1. The site excavators have argued that the presence of so many metates in Household 1 might point to periods of enhanced production for activities related to Structure 10 (the "feasting" structure).

The association of botanical remains in handled jars, the distribution of metates, and the close proximity of these objects provides a high-resolution perspective on how these activities might have been organized on the August evening prior to the Loma Caldera explosion. The presence of the duck, and the relatively object-free space around it, only reinforces the notion that spaces were not inert physical areas to be filled with lifeless objects but dynamic spaces where things imbued with value were arranged in meaningful ways. On display, on the one hand, is a remarkable snapshot of Cerén lifeways at a particular moment in time. However, also visible are taskscapes, comprised of the combined set and series of tasks, with each task taking meaning from a position within a broader ensemble and in relation to spatial entanglements (Ingold 2000).

GIS can be an invaluable tool in investigating these various facets of life at Cerén due to its ability to provide multi-scalar analyses of practice. It is clear from this case study that GIS analyses at the excavation-level can yield insights into daily life to complement studies on large-scale settlement and related phenomena. In combination, these approaches yield a more robust set of lifeways and a deeper view of household spatiality, providing a rich set of analogs for use at other Mesoamerican sites where preservation is less remarkable (Cf. Webster et al. 1997).

References

Andrienko, N., & Andrienko., G. (2006). *Exploratory analysis of spatial and temporal data: A systematic approach*. New York: Springer.

Ashmore, W., & Wilk, R. R. (1988). House and household in the Mesoamerican past: An introduction. In W. Ashmore & R. R. Wilk (Eds.), *Household and community in the Mesoamerican past* (pp. 1–28). Albuquerque, NM: University of New Mexico Press.

Beaudry-Corbett, M., & Bishop, R. L. (2002). Ceramics and their use at Cerén. In P. D. Sheets (Ed.), *Before the volcano erupted: The ancient Cerén village in Central America* (pp. 117–138). Austin, TX: University of Texas Press.

Beaudry-Corbett, M., Simmons, S. E., & Tucker, D. B. (2002). Ancient home and garden: The view from structure 1 at Cerén. In P. D. Sheets (Ed.), *Before the volcano erupted: The ancient Cerén village in Central America* (pp. 45–57). Austin, TX: University of Texas Press.

Berdan, F. F., & Anawalt, P. R. (1992). *The Codex Mendoza*. Berkeley, CA: University of California Press.

Boudreaux, E. A. (2007). *The archaeology of Town Creek*. Tuscaloosa, AL: University of Alabama Press.

Bray, T. L. (2003). Inka pottery as culinary equipment: Food, feasting and gender in imperial state design. *Latin American Antiquity, 14*(1), 3–28.

Church, T., Brandon, R. J., & Burgett, G. R. (2000). GIS applications in archaeology: Method in search of theory. In K. L. Wescott & R. J. Brandon (Eds.), *Practical applications of GIS for archaeologists: A predictive modeling toolkit* (pp. 144–159). London: Taylor & Francis.

Clark, J. E. (1988). *The lithic objects of La Libertad, Chiapas, Mexico: an economic perspective*. Provo, UT: New World Archaeological Foundation, Brigham Young University.

Coe, M. D., & Kerr, J. (1982). *Old gods and young heroes: The Pearlman collection of Maya ceramics*. Jerusalem: Israel Museum.

de Landa, D. (1978). *Yucatán before and after conquest* (W. Gates, Trans., *Relación de las cosas de Yucatán*, 1566). New York: Dover.

Fletcher, L. A. (1983). Coba and Mayapan: A comparison of solares, household variation, socioploitical organization, and land tenure. In W. Folan, E. Kintz, & L. A. Fletcher (Eds.), *Coba, a classic Maya metropolis* (pp. 121–131). New York: Academic Press.

Fotheringham, A. S., Brundsdon, C., & Charlton, M. (2000). *Quantitative geography: Perspectives on spatial data analysis*. London: SAGE.

Fuller, D. Q. (2005). Ceramics, seeds, and culinary change in prehistoric India. *Antiquity, 76*(306), 761–777.

Getis, A., & Ord, J. K. (1992). The analysis of spatial association by use of distance statistics. *Geographical Analysis, 24*(3), 189–206.

Golden, C. W. (2002). *Bridging the gap between archaeological and indigenous chronologies: An investigation of the early classic/late classic divide at Piedras Negras, Guatemala*. Ph.D Dissertation, University of Pennsylvania, Philadelphia, PA.

Hanks, W. (1990). *Referential practice: Language and lived space among the Maya*. Chicago, IL: University of Chicago Press.

Hartman, C. V. (1907). Archaeological researches on the Pacific Coast of Costa Rica. *Memoirs of the Carnegie Museum, 3*, 1–188.

Hayden, B. (1987). *Lithic studies among the contemporary highland Maya*. Tucson, AZ: University of Arizona Press.

Hendon, J. A. (1996). Archaeological approaches to the organization of domestic labor: Household practice and domestic relations. *Annual Review of Anthropology, 25*(1), 45–61.

Hendon, J. A. (1997). Women's work, women's space, and women's status among the Classic-period Maya elite of the Copan Valley, Honduras. In C. Claassen & R. A. Joyce (Eds.), *Women in prehistory: North America and Mesoamerica* (pp. 33–46). Philadelphia, PA: University of Pennsylvania Press.

Houston, S. D., Stuart, D., & Taube, K. A. (2006). *The memory of bones: Body, being, and experience among the classic Maya*. Austin, TX: University of Texas Press.

Ingold, T. (1993). The temporality of the landscape. *World Archaeology, 25*(2), 152–174.

Ingold, T. (2000). *The perception of the environment: Essays on livelihood, dwelling and skill*. London: Routledge.

Inomata, T., & Stiver, L. R. (1998). Floor assemblages from burned structures at Aguateca, Guatemala: A study of classic Maya households. *Journal of Field Archaeology, 25*(4), 431–452.

Lange, F. W., & Rydberg, C. R. (1972). Abandonment and post-abandonment behavior at a rural Central American house-site. *American Antiquity, 37*(3), 419–432.

Lentz, D. L., Beaudry-Corbett, M., Reyna de Aguilar, M. L., & Kaplan, L. (1996). Foodstuffs, forests, fields, and shelter: A paleoethnobotanical analysis of vessel contents from the Cerén Site. El Salvador. *Latin American Antiquity, 7*(3), 247–262.

Lentz, D. L., & Ramirez-Sosa, C. R. (2002). Cerén plant resources: Abundance and diversity. In P. D. Sheets (Ed.), *Before the volcano erupted* (pp. 33–42). Austin, TX: University of Texas Press.

Levi, L. J. (1996). Sustainable production and residential variation: A historical perspective on pre-Hispanic domestic economies in the Maya Lowlands. In S. L. Fedick (Ed.), *The managed mosaic: Ancient Maya Agriculture and resource use* (pp. 92–106). Salt Lake City: University of Utah Press.

Manzanilla, L., & Barba, L. (1990). The study of activities in classic households: Two case studies from Coba and Teotihuacan. *Ancient Mesoamerica, 1*(1), 41–49.

Mayer, C. (2006). Making use of distances: Estimating parameters of spatial processes. In M. W. Mehrer & K. L. Wescott (Eds.), *GIS and archaeological site location modeling* (pp. 137–153). London: Taylor & Francis.

McKee, B. R. (1999). Household archaeology and cultural formation processes: Examples from the Cerén Site, El Salvador. In P. M. Allison (Ed.), *The archaeology of household activities* (pp. 30–42). London: Routledge.

McKee, B. R. (2002). Radiocarbon dating and chronology. In P. D. Sheets (Ed.), *Before the volcano erupted: The ancient Cerén village in Central America* (pp. 7–8). Austin, TX: University of Texas Press.

Miller, C. D. (2002). Volcanology, stratigraphy, and effects on structures. In P. D. Sheets (Ed.), *Before the volcano erupted: The ancient Cerén village in Central America* (pp. 11–23). Austin, TX: University of Texas Press.

Moholy-Nagy, H. (2003). *The objects of Tikal: Utilitarian objects and unworked material* (Tikal Report No. 27, Part B). Philadelphia, PA: University of Pennsylvania Museum.

Montgomery, J. (2000). The Montgomery drawings collection. Foundation for the Advancement of Mesoamerican Studies, Inc. http://research.famsi.org/montgomery.html. Accessed 29 February 2016.

Neubauer, W. (2004). GIS in archaeology—The Interface between prospection and excavation. *Archaeological Prospection, 11*(3), 159–166.

Plunket, P., & Uruñuela, G. (2000). The quick and the dead: Decision-making in the abandonment of Tetimpa. *Mayab, 13,* 78–87.

Redfield, R. (1950). *A village that chose progress: Chan Kom revisited.* Chicago, IL: University of Chicago Press.

Redfield, R., & Villa Rojas, A. (1934). *Chan Kom: A Maya village.* Washington, D.C.: Carnegie Institution.

Schneider, J. (2002). Milling tool design, stone textures and function. In R. Treuil & H. Procopiou (Eds.), *Moudre et Broyer* (pp. 31–53). Paris: Publications du C.N.R.S.

Searcy, M. T. (2011). *The life-giving stone: Ethnoarchaeology of Maya Metates.* Tucson, AZ: University of Arizona Press.

Sheets, P. D. (1998). Place and time in activity area analysis: A study of elevated contexts used for object curation at the Cerén Site, El Salvador. *Revista Española de Antropología Americana, 28,* 63–98.

Sheets, P. D. (2000). Provisioning the Cerén household: The vertical economy, village economy, and household economy in the Southeastern Maya periphery. *Ancient Mesoamerica, 11*(2), 217–230.

Sheets, P. D. (Ed.). (2002a). *Before the volcano erupted: The ancient Cerén village in Central America.* Austin, TX: University of Texas Press.

Sheets, P. D. (2002b). Groundstone objects in the Cerén Village. In P. D. Sheets (Ed.), *Before the volcano erupted: The ancient Cerén village in Central America* (pp. 145–150). Austin, TX: University of Texas Press.

Sheets, P. D., & Brown, L. A. (Eds.). (1996). *Preliminary report of the Cerén research project, 1996 field season. Prepared Report.* Boulder, CO: Department of Anthropology, University of Colorado.

Sheets, P. D., & Kievet, K. A. (Eds.). (1992). *1992 archaeological investigations at the Cerén Site, El Salvador: A preliminary report. Report prepared for the National Science Foundation.* Boulder, CO: Department of Anthropology, University of Colorado.

Sheets, P. D., & McKee, B. R. (Eds.). (1989). *1989 archaeological investigations at the Cerén Site, El Salvador: A preliminary report. Report prepared for the National Science Foundation.* Boulder, CO: Department of Anthropology, University of Colorado.

Sheets, P. D., & McKee, B. R. (Eds.). (1990). *1990 archaeological investigations at the Cerén Site, El Salvador: A preliminary report. Report prepared for the National Science Foundation.* Boulder, CO: Department of Anthropology, University of Colorado.

Sheets, P. D., & Simmons, S. E. (Eds.). (1993). *Preliminary report of the Cerén Research Project, 1993 Season. Prepared report.* Boulder, CO: Department of Anthropology, University of Colorado.

Sheets, P. D., & Simmons, S. E. (2002). Household production and specialization at Cerén. In P. D. Sheets (Ed.), *Before the volcano erupted: The ancient Cerén village in Central America* (pp. 178–183). Austin, TX: University of Texas Press.

Sheets, P. D., & Woodward, M. (2002). Cultivating biodiversity: Milpas, gardens, and the classic period landscape. In P. D. Sheets (Ed.), *Before the volcano erupted: The ancient Cerén village in Central America* (pp. 184–191). Austin, TX: University of Texas Press.

Siart, C., Eitel, B., & Panagiotopoulos, D. (2008). Investigation of past archaeological landscapes using remote sensing and GIS: A multi-method case study from Mount Ida. *Crete. Journal of Archaeological Science, 35*(11), 2918–2926.

Sweely, T. L. (1998). Personal interactions: The implications of spatial arrangements for power relations at Cerén. El Salvador. *World Archaeology, 29*(3), 393–406.

Taliaferro, M. S., Schriever, B. A., & Shackley, M. S. (2010). Obsidian procurement, least cost path analysis, and social interaction in the Mimbres area of Southwestern New Mexico. *Journal of Archaeological Science, 37*(3), 536–548.

Van Derwarker, A., Alvarado, J. V., & Webb, P. (2014). Analysis and interpretation of intrasite variability in paleoethnobotanical remains: A consideration and application of methods at the Ravensford Site, North Carolina. In J. M. Marston, & J. d'Alpoim Guedes (Eds.), *Method and theory in paleoethnobotany* (pp. 205–234). Boulder, CO: University Press of Colorado.

Ward, J. H., Jr. (1963). Hierarchical grouping to optimize an objective function. *Journal of the American Statistical Association, 58*(301), 236–244.

Wauchope, R. (1938). *Modern Maya houses: A study of their archaeological significance.* Washington, D.C.: Carnegie Institution of Washington.

Webster, D., & Gonlin, N. (1988). Household remains of the humblest Maya. *Journal of Field Archaeology, 15*(2), 169–190.

Webster, D., Gonlin, N., & Sheets, P. D. (1997). Copan and Cerén. *Ancient Mesoamerica, 8*(1), 43–61.

Wilk, R. R., & Rathje, W. (1982). Archaeology of the household: Building a prehistory of domestic life. *American Behavioral Scientist, 25*(2), 611–724.

Wilson, G. D. (2008). *The archaeology of everyday life at early Moundville.* Tuscaloosa, AL: University of Alabama Press.

Chapter 6
Ritual and Plant Use at Conchopata: An Andean Middle Horizon Site

Matthew P. Sayre and William T. Whitehead

Introduction

> Andean ritual is materialized in architecture, suggesting procession and pilgrimage, and in hearths and ceramics, informing us about smell, taste, hearing, touch, and sight—and thus about smoke and alcoholic transformation, marking the shift from real to ritual time and demonstrating the powers of the telluric ancestors and the life-giving forces of the dead. Several theoretical approaches to ritual exist within the Andean literature, but there is much that can still be studied with these rich data sets that illustrate a range of rituals. (Hastorf 2007: 97)

The corporeal elements of ritual benefit from and are enhanced by plants. The food, drink, and mind-altering substances of the ancient world were critical elements of past ritual practices. Plants can be used in myriad manners, but in many instances plant-derived products are linked to specific occasions. For example, sacramental wine and table wine are both made from grapes, yet the spaces in which they are consumed and their significance could not be more different. The separation of the ritual object or event from its quotidian counterpart is a constant challenge in archaeology. This is especially true when the same spaces and objects are the loci of these events. Here we attempt to shed light on some of these tensions in our analysis of past plant use through the analysis of spatial, architectural, ceramic, and botanical evidence. This chapter will examine past ritual practice at Conchopata, a Middle Horizon (650–1000 AD) site in Peru. The results from the site will be compared to those from other Middle Horizon sites. We also examine the initial

M.P. Sayre (✉)
Department of Anthropology, University of South Dakota,
326 East Hall, Vermillion, SD 57069, USA
e-mail: Matthew.Sayre@usd.edu

W.T. Whitehead
SWCA Environmental Consultants,
5647 Jefferson Street NE, Albuquerque, NM 87109, USA
e-mail: wwhitehead@swca.com

M.P. Sayre and M.C. Bruno (eds.), *Social Perspectives on Ancient Lives from Paleoethnobotanical Data*, DOI 10.1007/978-3-319-52849-6_6

architectural designations of ritual and domestic space at the site and consider how botanical data can critically inform this separation of use areas.

Time and Place

The Middle Horizon witnessed many changes in the scale of social organization and trade, as well as the beginnings of the archaic states of Wari and Tiwanaku in the Andes (Isbell and Cook 2002; Schreiber 1992). However large or grandiose these phenomena became, both Wari and Tiwanaku emerged from smaller cultural antecedents (Huarpa and Late Formative Period cultures, respectively). Many researchers have studied the Wari phenomenon and associated Wari sites (Browman 1976; Earle and Jennings 2013; Glowacki 2002; Isbell 2004; Isbell and McEwan 1991; Isbell and Schreiber 1978; Jennings and Craig 2001; Kaulicke and Isbell 2000, 2001; McEwan 1990, 1996, 1998; Ochatoma Paravicino and Cabrera Romero 2000; Rowe et al. 1950; Schreiber 1992; Van Buren 1992; Williams 2001, 2002; Williams and Nash 2002). Unfortunately, very few Huarpa to Wari sites from the Early Intermediate Period and the subsequent Early Middle Horizon have been excavated or studied from a paleoethnobotanical perspective (Isbell 2008). Therefore, our understanding of botanical evidence from Conchopata, an important Wari center established two centuries before the rise of the Wari Empire, is vital to understanding how the early inhabitants of these Wari sites used plants and interacted with its environment. We also integrate botanical data with other forms of archaeological data, an approach that has proven useful in many areas of the world (e.g., Kintigh and Altschul 2010; Spielmann and Kintigh 2011; VanDerwarker and Peres 2010).

Conchopata lies approximately 10 km south of its neighboring contemporary site of Wari (Isbell (2008) refers to Conchopata as Ayacucho's "second city"). Much of the site has been destroyed, but originally it likely covered 20–40 ha. (Isbell 2008; Isbell and Cook 2002) (Figs. 6.1 and 6.2). Based on similar pottery styles, characteristic Wari orthogonal architecture, and its close proximity to the largest Wari site yet discovered, Conchopata has been classified as a Wari administrative site (Cook and Isbell 2000). Wari lifestyle, including certain ceramic and architectural styles, developed out of the community at Conchopata (Ochatoma Paravicino and Cabrera Romero 2002). Conchopata was occupied throughout Wari's rise as the capital of the Wari Empire, and has become a well-studied site (Cook and Benco 2000; Green and Whitehead 2006; Isbell 2008; Isbell and Cook 2002; Knobloch 2000; Lumbreras 1974; Ochatoma Paravicino and Cabrera Romero 2000; Pozzi-Escot 1991; Rosenfeld 2012; Sayre et al. 2012; Tung 2007, 2008; Tung and Cook 2006).[1]

---

[1]Ñawinpukyo, near Conchopata has also been studied extensively (Leoni 2005, 2006; Machaca 1997).

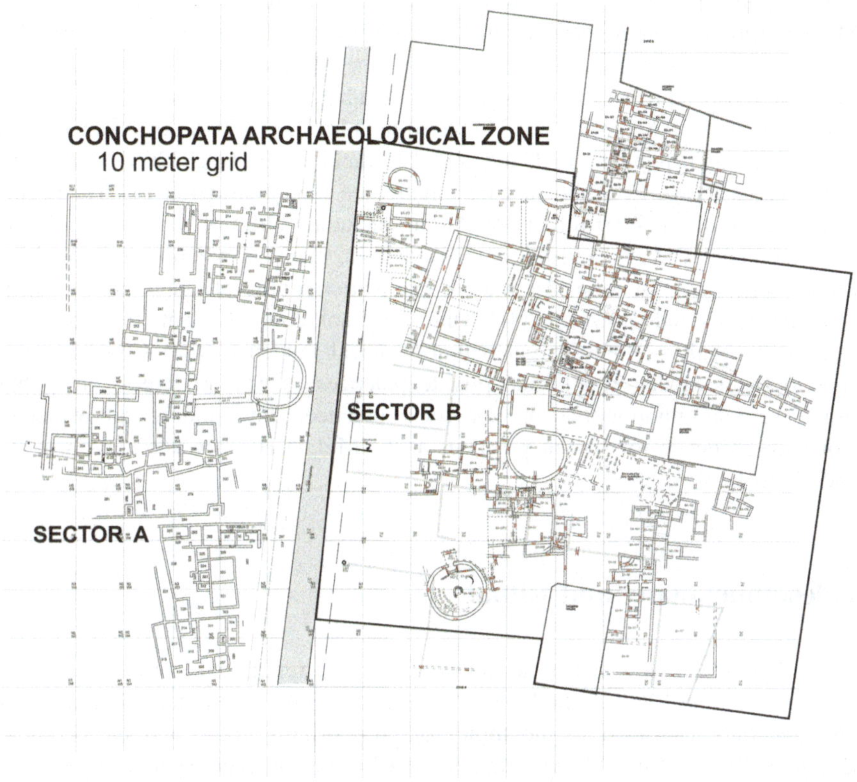

Fig. 6.1 Map of Conchopata showing all exposed architecture

Fig. 6.2 Conchopata during excavation

Conchopata was a densely occupied settlement of rectangular stone and mud structures, with occasional D-shaped structures, patio groups (also known as open plazas), and streets (Ochatoma Paravicino and Cabrera Romero 2002). The prevalence of ceramics and ceramic production tools at the site, especially in the site's abundant trash deposits, led to the original classification of Conchopata as a Middle Horizon ceramic production center occupied by an emerging middle class of craft specialists (Menzel 1964). However, the presence of several patio groups associated with elite tombs, in addition to evidence for large-scale feasting and other elite facilities (e.g., such as D-shaped temples) has shifted our understanding of Conchopata's population (Cook and Glowacki 2003; Isbell 2009; Rosenfeld 2012; Tung 2008). Conchopata's inhabitants are now considered to have been a cross-section of the Wari urban population, including elites of different rank, artisans, common workers, and religious specialists, many of whom may have been involved in the production, decoration, or commissioning of elaborate, oversized ceramics (Cook and Glowacki 2003; Tung and Cook 2006). Keeping these previous studies in mind, we present paleoethnobotanical evidence from Conchopata to support the idea that small-scale private practices as well as large-scale public rituals were an important part of Wari social practice. These findings are presented alongside evidence for similar practices at the Middle Horizon sites of Tiwanaku and Cerro Baúl.

Paleoethnobotany and Ritual

Paleoethnobotany as a discipline did not become a regular part of archaeological research until the 1960s in the United States and Europe, and at that point in time it was generally restricted to the topics of crop origins, food, and paleoecology (Renfrew 1973; Yarnell 1970). There has been a shift in recent years to expand the discussion of plants in the past to include more questions about the role of plants in social life (see Chiou et al. 2014; Hastorf 1999; Morehart and Morell-Hart 2013).

These later paleoethnobotanical analyses revealed patterns in ancient subsistence and economics, forming step two on Hawkes' ladder of inference (1954: 161). Hawke's well-known four-step ladder consists of the following rungs: (1) production techniques, (2) subsistence-economics, (3) social/political institutions, (4) religious institutions/spiritual life (Hawkes 1954: 161). Each ascending step is considered to be more difficult to analyze than the previous one. The ideational or philosophical realm is still considered less accessible than nutrition or threshing techniques. However, the interpretation of symbols may not be more removed from our comprehension than the interpretation of food remains. Plants can, and generally are, entangled in all levels of society and in many instances plants occupy critical roles in religious practice and discourse (Simpson and Conner-Ogorzaly 2013). For example, plants change behavioral practice when they are consumed for their consciousness-altering qualities. In many ancient societies, plants capable of producing visions were used in medicinal and/or religious practice (Sayre 2014; Wilbert 1993). While all plants can be toxins, those with the special ability to temporarily alter mental states can acquire particular resonance and power. These plants may have been considered as conduits to the gods, or as gods in their own right (Schultes and Hoffman 2001). Here, we attempt to use botanical data to reexamine the initial architectural designations (i.e. constructed for ritual or domestic use), thus mixing plant analysis into all four of Hawkes' ladder steps.

Domestic and Ritual Space

The distinction between ritual space and domestic space in modern built environments is not always evident to outsiders. This difficulty in accurate determination is only compounded when there is not only cultural difference to take into account but also temporal and taphonomic factors to consider. However, the variation in built structures does seemingly imply distinct use patterns. Moore laid out one useful set of criteria for examining Andean ritual architecture (1996: 139–165); his five categories were: permanence, centrality, ubiquity, scale, and visibility. These criteria are useful for understanding explicitly ritual architecture in the modern and ancient built environment. Examining these variables in a population of structures allows researchers to understand the importance or relative significance of buildings in relationship to the rest of the built environment. At Conchopata, domestic spaces were common and generally consisted of smaller rooms and dwelling areas. The ritual spaces were predominately D-shaped structures (commonly containing artifacts associated with religious activity), public spaces such as patios, and areas where offerings were encountered. The D-shaped structures at Wari, in the Vegachoyuc Moqo sector, were initially defined as ritual structures due to their high concentrations of decorated ceramics and presence of human burials or trophy heads (Bragayrac 1991). The D-shaped structures at Conchopata have been defined as having a similar function (Cook 2001). Of Moore's five criteria, the D-shaped structures and large patios appear to have been permanent, central, and ubiquitous. Their scale and visibility could be constrained by surrounding buildings, thus indicating that they were often located in or near domestic settlements.

Moore's five criteria for sacred architecture may be less useful for identifying architectural space that served *both* ritual and domestic purposes. This point is important as much of ritual life and the experience of the sacred can occur outside of formal architecture. The domestic spaces and related intermediary spaces where public events occurred are the sites of intimate repeated acts, engaging with worlds and powers beyond the mundane (Dean and Kojan 2001; Hastorf 2001).

The questions we raise are how does the paleoethnobotanical record at Conchopata interact with the schema of ritual versus domestic architecture, and how could the botanical data lead us to interpretations that may be statistically valid but culturally erroneous? The paleoethnobotanical samples presented for analysis were coded in an archaeologically constructed binary system of contextual identification. Some of them were labeled as being from domestic spaces and others were labeled as ritual space. The classification system was primarily defined on the basis of architectural form, such as patios or D-shaped structures, and contextually defined by the materials, such as burials or ceramics, which were uncovered over the course of excavations. These more contextual definitions of ritual space take into account a variety of criteria, such as artifacts and analytical evidence to understand how a space was used. To this list of criteria we add botanical remains, which reveal evidence of past meals, beverages, and burnt offerings. The patterning of botanical remains presents an opportunity to reconsider spatial and activity-use designations.

Previous research at Conchopata has identified distinct aspects of Wari ritual behavior. The architecture of the site clearly contains separate spaces for ritual activities, such as patios and D-Shaped structures (Isbell and Cook 2002). Ceramic analysis revealed that there were differences in vessel types depending upon the architectural space (Cook 2001; Cook and Glowacki 2003). Zooarchaeological analysis shows evidence of offerings of guinea pigs and juvenile camelids under house floors and alongside human remains (Rosenfeld 2012). The faunal analysis also shows the residues of large meals involving great quantities of camelid meat, in both public and private spaces associated with decorated ceramics (Rosenfeld 2012). The human remains recovered at the site provide insights into past daily life and ritual activities. Human burials were found in areas defined as both domestic and ritual and there was ample evidence to support the idea that many tombs were opened so that offerings could be deposited with the bodies (Isbell 2004). However, the presence of trophy heads found in the D-shaped structures indicate that the preserved remains of past enemies and/or ancestors were guarded and maintained in spaces separated from daily activities (Tung 2008).

Paleoethnobotany in the Middle Horizon

Methods

Conchopata was excavated between the years 1997 and 2003 (Cook and Isbell 2000; Isbell and Cook 2002; Ochatoma Paravicino and Cabrera Romero 2002). After 1998, one flotation sample was collected for each locus (unique archaeological context) that was excavated. The samples were processed using water flotation to collect carbon samples in the field (Pearsall 2000). Materials were initially sorted in the field in order to discern the range of preservation conditions and materials present. Then we, M. Sayre, W. Whitehead, and K. Green, sorted the materials into identification categories at the University of California at Berkeley and Ripon College. The volumes of soil samples were recorded so that we could assess standardized density of plant remains, and the cultural context and archaeological provenience information allowed us to examine ubiquity (percent presence) across different spaces (Popper 1989).

Comparative Paleoethnobotany

The paleoethnobotanical exploration of Wari sites is only now beginning to bear fruit with recent analyses for the Moquegua Valley sites of Cerro Baúl and Cerro

Trapiche (Goldstein and Coleman 2004; Green and Whitehead 2006; Parrish 1999; Sayre et al. 2012; Sayre and Whitehead 2002). There has been a longer tradition of paleoethnobotanical studies in the Lake Titicaca Basin of the Andes, particularly for the other major Middle Horizon state of Tiwanaku and its primary urban center located in the highlands of Bolivia (Kolata 1993; Wright et al. 2003). In order to facilitate the interpretation of the Conchapata plant remains, we compare our findings to those from these roughly contemporaneous sites. Detailed data on a range of taxa have been published for Tiwanaku (Wright et al. 2003), so we include in our analysis a more detailed comparison with that site. We then turn to broader comparisons of plant remains as they related to the brewing of *chicha* at Cerro Baúl.

Results

A total of 32,234 specimens from Conchopata were examined and placed in 49 taxonomic categories. These categories included 26 families, 24 genera, and 8 species-level identifications (see Summary Table 6.1). They are separated by screen size to facilitate future comparisons of change in taxa size over time. Below, we discuss in detail the plant taxa that are most common, but that also relate most closely to domestic and ritual practices, such as food, drink, and stimulant use. We then compare their patterns to those at Tiwanaku (Wright et al. 2003).

Chenopodium quinoa was the most common plant type found at Conchopata, yielding both high density and ubiquity values (see Tables 6.1 and 6.2). *Chenopodium* seeds were found in almost every context, whether ritual or domestic. The median density per liter was 0.4 seeds with 37 architectural units falling above this mark. The majority of contexts containing *Chenopodium* (73%) were interpreted as domestic in nature. This distribution may indicate several likely uses of *Chenopodium*: the seeds were used for food, stalks and stems were used for fuel, and the alkaline ash from the burning of quinoa could have been used as an activator lime during coca chewing.

At Conchopata we have a snapshot of possible selection for larger seeded quinoa during the Middle Horizon (Fig. 6.3). Our analysis shows that the majority of *Chenopodium* seeds found at Conchopata were within the size range of modern domesticated quinoa (*Chenopodium quinoa* L). varieties (2.95–1.05 mm) (Bruno 2006: 39); however, there was still a high proportion of small, weedy varieties (2.00–<0.5 mm) (Table 6.1). This is larger than seeds from the Formative period (1500–800 BC) site of Chiripa (Bruno and Whitehead 2003).

At the site of Tiwanaku, *Chenopodium* was not commonly identified in ceremonial areas, but it was common in domestic spaces (Wright et al. 2003). *Chenopodium,* while present in some elite households, was particularly ubiquitous at the secondary center of Lukurmata than at Tiwanaku (Wright et al. 2003: 390).

Schinus molle: The Andean pepper tree (*Schinus molle*) is a versatile plant resource used not only for fuel and construction, but also for the brewing of a strong *chicha* (Goldstein and Coleman 2004; Moseley et al. 2005; Sayre et al. 2012).

Table 6.1 All identification categories (density and ubiquity)

0.5 mm screen					>1 mm screen				
Identification Category	Total	Total Loci	0.5 mm Density	0.5 mm ubiquity	Identification Category	Total	Total Loci	1 mm Density	1 mm Ubiquity
Amaranthaceae – *Amaranthus* sp.	12	7	0.01	0.04	Amaranthaceae – *Amaranthus* sp.	1	1	0.00	0.00
Anacardiaceae – *Schinus molle*	196	35	0.12	0.21	Anacardiaceae – *Schinus molle*	2306	175	0.55	0.42
Apiaceae	2	1	0.00	0.01	Apiaceae			0.00	0.00
Asteraceae	10	8	0.01	0.05	Asteraceae	14	5	0.00	0.01
Asteraceae – *Ambrosia* sp.			0.00	0.00	Asteraceae – *Ambrosia* sp.	2	2	0.00	0.00
Boraganaceae	113	7	0.07	0.04	Boraganaceae			0.00	0.00
Brassicaceae	10	7	0.01	0.04	Brassicaceae	4	4	0.00	0.01
Cactaceae	10	9	0.01	0.05	Cactaceae	24	16	0.01	0.04
Cactaceae – *Cerus* sp.	1	1	0.00	0.01	Cactaceae – *Cerus* sp.			0.00	0.00
Cactaceae – Opuncioideae	2	1	0.00	0.01	Cactaceae – Opuncioideae			0.00	0.00
Capirfoliacae			0.00	0.00	Capirfoliacae	4	1	0.00	0.00
Chenopodiaceae – *Chenopodium* sp.			0.00	0.00	Chenopodiaceae – *Chenopodium* sp.	13	1	0.00	0.00
Chenopodiaceae – *Chenopodium quinoa*	2169	120	1.32	0.73	Chenopodiaceae – *Chenopodium quinoa*	4239	287	1.01	0.68
Cyperaceae	31	21	0.02	0.13	Cyperaceae	6	3	0.00	0.01
Cyperaceae – *Scirpus* sp.	8	2	0.00	0.01	Cyperaceae – *Scirpus* sp.	3	2	0.00	0.00
Dung	44	5	0.03	0.03	Dung	73	24	0.02	0.06
Erythroxiaceae – *Erythroxylum coca*	7	1	0.00	0.01	Erythroxiaceae – *Erythroxylum coca*	2	1	0.00	0.00
Fabaceae	30	13	0.02	0.08	Fabaceae	187	79	0.04	0.19

(continued)

Table 6.1 (continued)

0.5 mm screen					>1 mm screen				
Identification Category	Total	Total Loci	0.5 mm Density	0.5 mm ubiquity	Identification Category	Total	Total Loci	1 mm Density	1 mm Ubiquity
Fabaceae – *Acacia* sp.			0.00	0.00	Fabaceae – *Acacia* sp.	1	1	0.00	0.00
Fabaceae – Mimosoideae			0.00	0.00	Fabaceae – Mimosoideae	2	1	0.00	0.00
Fabaceae – *Phaseolus lunatis*			0.00	0.00	Fabaceae – *Phaseolus lunatis*	2	2	0.00	0.00
Fabaceae – *Phaseolus* sp.	9	2	0.01	0.01	Fabaceae – *Phaseolus* sp.	8	5	0.00	0.01
Fabaceae – *Phaseolus vulgaris*			0.00	0.00	Fabaceae – *Phaseolus vulgaris*	1	1	0.00	0.00
Juncaceae			0.00	0.00	Juncaceae	9	1	0.00	0.00
Lamiaceae	13	8	0.01	0.05	Lamiaceae	15	8	0.00	0.02
Malvaceae	185	42	0.11	0.26	Malvaceae	89	21	0.02	0.05
Molluginaceae – *Mollugo* sp.	7	1	0.00	0.01	Molluginaceae – *Mollugo* sp.	1	1	0.00	0.00
Nyctaginaceae	1	1	0.00	0.01	Nyctaginaceae			0.00	0.00
Parenchyma	469	74	0.29	0.45	Parenchyma	1915	249	0.46	0.59
Poaceae	103	49	0.06	0.30	Poaceae	108	15	0.03	0.04
Poaceae – Panicoid			0.00	0.00	Poaceae – Panicoid	1	1	0.00	0.00
Poaceae – *Stipa ichu*			0.00	0.00	Poaceae – *Stipa ichu*	4	4	0.00	0.01
Poaceae – *Zea mays*	114	43	0.07	0.26	Poaceae – *Zea mays*	1769	361	0.42	0.86
Polygonaceae – *Polygonum* sp.	5	2	0.00	0.01	Polygonaceae – *Polygonum* sp.	3	3	0.00	0.01
Polypodiaceae	2	1	0.00	0.01	Polypodiaceae			0.00	0.00
Portulacaceae – *Portulaca* sp.	134	41	0.08	0.25	Portulacaceae – *Portulaca* sp.	14	5	0.00	0.01
Rannunculaceae			0.00	0.00	Rannunculaceae	1	1	0.00	0.00
Rubiaceae – *Galium* sp.			0.00	0.00	Rubiaceae – *Galium* sp.	2	2	0.00	0.00

(continued)

Table 6.1 (continued)

0.5 mm screen					>1 mm screen				
Identification Category	Total	Total Loci	0.5 mm Density	0.5 mm ubiquity	Identification Category	Total	Total Loci	1 mm Density	1 mm Ubiquity
Rubiaceae – *Relbunium* sp.	11	9	0.01	0.05	Rubiaceae – *Relbunium* sp.	1	1	0.00	0.00
Rubiaceae – *Rubus* sp.			0.00	0.00	Rubiaceae – *Rubus* sp.	1	1	0.00	0.00
Solanaceae	132	10	0.08	0.06	Solanaceae			0.00	0.00
Solanaceae – *Capsicum* sp.			0.00	0.00	Solanaceae – *Capsicum* sp.	1	1	0.00	0.00
Solanaceae – *Nicotiana* sp.	4	4	0.00	0.02	Solanaceae – *Nicotiana* sp.			0.00	0.00
Solanaceae – *Solanum* sp.	324	55	0.20	0.34	Solanaceae – *Solanum* sp.	15	10	0.00	0.02
Syceringium	1	1	0.00	0.01	Syceringium			0.00	0.00
Unidentifiable	841	68	0.51	0.41	Unidentifiable	320	52	0.08	0.12
Unidentified	153	33	0.09	0.20	Unidentified	239	117	0.06	0.28
Verbenaceae – *Verbena* sp.	86	2	0.05	0.01	Verbenaceae – *Verbena* sp.	56	1	0.01	0.00
Grand total	5239				Grand total	11456			

Table 6.2 Areas of Highest Chenopod Concentration

Species	*Chenopodium quinoa*			
Area	Total specimens	Total loci	Density seeds per liter	Ubiquity per loci
A186	368	4	9.20	1.00
A180	1057	10	9.61	0.91
A174 T1	338	4	8.45	1.00
A176 T1	187	3	6.23	1.00
A174	39	1	3.90	1.00
A149	88	5	1.76	1.00
A6	174	4	3.48	0.80
A133	136	10	1.36	1.00
A160	13	1	1.30	1.00
A182	189	7	2.36	0.88
A116	104	11	0.95	1.00
A188	55	6	0.92	1.00
A164	9	1	0.90	1.00

Fig. 6.3 Example of *Chenopodium quinoa* seed

These seeds (not a food source consumed by humans) were the third most common species at Conchopata by density and second most abundant by ubiquity (see Table 6.3 and Fig. 6.4). The top ten contexts for *molle* show a slight bias toward ritual contexts (60%), which aligns well with the common interpretation that *molle* was a ritual plant (Moseley et al. 2005). Since the fruit and seeds themselves are not consumed directly but processed as part of an alcoholic brew, their elevated density and ubiquity are strong indicators that molle enjoyed special status. *Schinus molle* was not initially recovered at the site of Tiwanaku (Wright et al. 2003); this is likely because the tree's environmental range does not extend into the Bolivian *altiplano*. However, recent work in the sector Mollo Kontu of Tiwanaku has uncovered two molle seeds (Bruno and Ramos 2009).

Parenchyma and Tubers: Parenchyma is a diagnostic category used to describe all material remains that come from storage tissue in plants. The primary types of plants

Table 6.3 Areas of highest *Schinus molle* concentration

Species	*Schinus molle*			
Area	Total specimens	Total loci	Density seeds per liter	Ubiquity per loci
A40 B	524	6	7.49	0.86
A174 T1	286	3	7.15	0.75
A192	258	6	3.69	0.86
A143 T3	67	3	2.23	1.00
A180	299	8	2.72	0.73
A149	32	5	0.64	1.00
A132	5	1	0.50	1.00
A174	5	1	0.50	1.00
A164	4	1	0.40	1.00
A172	10	3	0.33	1.00
A143 T2	121	13	0.86	0.93
A160	3	1	0.30	1.00

Fig. 6.4 Example of *Schinus molle* seed

that have this kind of dense storage tissue in the Andes are tubers, especially the white potato (*Solanum tuberosum*). Table 6.4 shows the distribution for the top 10 densest areas containing parenchyma. The median density per liter was 0.24 objects per liter, with 35 architectural units falling above this mark, and 57% of these contexts were interpreted as domestic. This distribution was very similar to other plant remains discussed so far: parenchyma was found in small domestic rooms and areas adjacent to patios and ritual spaces. Therefore, we can safely conclude that the preparation, cooking, and consumption occurred in or near these areas.

In contrast, tuber remains were rare at Tiwanaku although they were recovered in different densities depending on the context (Wright et al. 2003: 390). The center of the site at Tiwanaku, rather than the domestic area of the site, Lukurmata, contained the greatest density of tuber remains, and they were associated with discrete ritual contexts, which led the investigators to the conclusion that they were an important part of ritual events (Wright et al. 2003: 392).

Table 6.4 Areas of Highest Parenchyma Concentration

Species	Parenchyma			
Area	Total specimens	Total loci	Density fragments per liter	Ubiquity per loci
A88	564.00	13.00	2.56	0.59
A186	47.00	4.00	1.18	1.00
A180	116.00	11.00	1.05	1.00
A93	37.00	4.00	0.93	1.00
A152 T1	8.00	1.00	0.80	1.00
A189	30.00	4.00	0.75	1.00
A44 A	13.00	2.00	0.65	1.00
A37	6.00	1.00	0.60	1.00
A153	46.00	3.00	1.15	0.75
A192	80.00	5.00	1.14	0.71

Table 6.5 Highest areas with *Zea mays* Cupule concentration

Species	*Zea mays*	Cupule		
Area	Total specimens	Total loci	Density cupules per liter	Ubiquity per loci
A176 T2	36	1	3.60	1.00
A176 T1	95	3	3.17	1.00
A172	11	5	0.37	1.67
A24 T1	9	4	0.30	1.33
A186	44	4	1.10	1.00
A172 T1	32	3	1.07	1.00
A112 T7	9	1	0.90	1.00
A197	6	1	0.60	1.00
A174 T1	22	4	0.55	1.00
A174	5	1	0.50	1.00
A37	5	1	0.50	1.00

Zea mays: Maize is a versatile plant, consumed as solid food by humans and domestic animals or brewed into the alcoholic beverage *chicha* (Hastorf 1990; Jennings and Bowser 2008; Moore 1989). Maize was common and abundant at Conchopata (see Tables 6.5 and 6.6 for cupule and kernel high density areas), but not as prevalent as *Chenopodium* or *Schinus molle*. There was a fairly even split between domestic and ritual areas in the presence of maize but the higher concentrations occurred in domestic contexts (eight of the top ten areas were domestic areas). The median cupule density was 0.14, and 31 contexts produced cupule densities above this mark. The split between ritual and domestic areas was also roughly even, but three areas with the greatest density of remains were domestic and were dramatically denser than other ritual or domestic areas. The kernel to cupule ratio was 1.16 (slightly more kernels to cupule fragments). This nearly even ratio suggests that maize was brought on the cob to the site and processed locally

Table 6.6 Areas of highest *Zea mays* Kernel concentration

Species	*Zea mays*	Kernel		
Area	Total specimens	Total loci	Density kernels per liter	Ubiquity per loci
A174	20	1	2.00	1.00
A176 T1	37	3	1.23	1.00
A40 B	105	5	1.50	0.71
A158	14	2	0.70	1.00
A152 T1	5	1	0.50	1.00
A37	5	1	0.50	1.00
A192	107	3	1.53	0.43
A170	41	6	0.59	0.86
A174 T2	8	3	0.27	1.00
A130	2	1	0.20	1.00
A197	2	1	0.20	1.00

Fig. 6.5 Example of carbonized *Zea mays* cupules

(Fig. 6.5). The maize from Conchopata was a small many-rowed variety; no kernels were greater than 4 mm in any dimension although the majority of recovered remains were fractured, and the cupules were equally small.

The possibility that maize was primarily used for *chicha* is in part supported by the presence of immense vessels (125–140 cm in height, over 90 L capacities) found near ceremonial spaces and likely used to produce this beverage for feasts at the site (Isbell 2008, 2009). There is additional isotopic evidence that demonstrates maize was a staple crop at the site (Finucane et al. 2006).

At the site of Tiwanaku, maize was not commonly found in ceremonial areas, but it was present in domestic areas and was found in higher densities in elite residences than in more humble residences (Wright et al. 2003: 397). It was imported from lower altitude areas because it cannot be grown where Tiwanaku is located in the *altiplano* (Hastorf et al. 2006).

Fig. 6.6 Carbonized Coca fruit

Erythroxylum coca: Coca is a plant synonymous with Andean life and is presumed to have a long history of use in South America (Plowman 1984, 1985). By far, the majority of paleoethnobotanical coca finds have been at coastal sites where preservation is excellent (Cartmell et al. 1991; Knudson and Buikstra 2007). Coca leaves may be found in mummy bundles, or with burials, while indirect evidence for coca use on the coast and in the highlands consists of paraphernalia for (and art showing) its use (Aufderheide et al. 1991; Cartmell et al. 1991; Kolata 1993; Lanning 1967; Lumbreras 1974; Moseley 1992). We recovered nine coca seed specimens, most likely early *E. coca* var. *coca* seeds (Fig. 6.6). All of these seeds were found in EA-148, a small domestic room in the east of the area excavated. Because the leaves, rather than the seeds, are chewed, seeds are unusual finds in botanical remains and leaves do not typically preserve as they are rarely charred. The recovery of charred coca seeds on site indicates that large portions of the plant were available and processed at Conchopata. In contrast, coca was not recovered at Tiwanakau, but it has been assumed to have been imported to the site (Wright et al. 2003: 402).

Comparisons to Cerro Baúl: Evidence for Wari Brewing and Feasting

The paleoethnobotany of Wari sites is still in its early stages. The most thoroughly analyzed site (paleoethnobotanically) and the best comparison to Conchopata is Cerro Baúl (Moseley et al. 2005) particularly as it relates to evidence for brewing and feasting, activities that would have been central to ritual behaviors. At Cerro Baúl, large quantities of *Schinus molle* and *Zea mays* as well as brewing and serving ceramic wares, *keros*, and other implements (such as grinding stones and ash pits) were recovered from rooms in the brewery complex (Goldstein and Coleman 2004; Moseley et al. 2005; Sayre et al. 2012). The north boiling room contained at least seven boiling pits along with opposed stone pedestals (Moseley et al. 2005: 17267). This led the archaeologists to argue for the presence of a brewing complex at this important Wari site.

At Conchopata, we also find architectural units with direct evidence for brewing and drinking including ceramics such as oversized, ornate ritual urns and copious amounts of individual drinking vessels. As at Cerro Baul, there appear to be separate rooms devoted to the production of large quantities of food and drink and separate public spaces for serving communal food. While we now have botanical evidence of plants used in brewing alcoholic beverages, there is weak botanical evidence of feasting preparation. There is mixed evidence of multi-functional site spaces more common in domestic sites.

Clearly, Cerro Baúl has the best evidence as yet for ritual feasting and brewing, but Conchopata was a different type of site with a longer occupational sequence. Conchopata was larger in size, clearly a ritual/administrative center, and seems to have supported a more diverse set of activities. This diversity of domestic activities likely led to the complex and multifaceted botanical record. As an early expression of Wari, Conchopata demonstrated patterns of *chicha* making using multiple species. This comparison makes clear that domestic and ritual space did not have consistent use patterns at Wari sites; rather, the nature of the site itself greatly impacted the relative segregation of domestic and ritual space.

Architecture and Plants: Domestic and Ritual Space

Based on the analysis presented above, we agree with Isbell (2008) that there does not appear to have been a clear separation between ritual and domestic space at Conchopata. The archaeological record contains evidence that multiple activities were conducted and multiple artifact types were abundant in both ritual and domestic areas (Isbell 2008, 2009). Table 6.7 shows that the contexts with the top 10 combined plant densities at Conchopata include six domestic and four ritual areas. The remainder of the combined plant densities does not follow a clear spatial pattern across ritual or domestic space. This lack of clear patterning is similar to other sites in the highlands. Hastorf (2001) and Dean and Kojan (2001) explore a similar situation at the site of Chiripa, near Lake Titicaca, discussing the importance of blurring the boundaries between ritual and domestic activities and loosening the theoretical and methodological systems we use for defining and studying these archaeological areas.

The map of Conchopata (see Fig. 6.1) shows a number of distinct areas that could be considered local group spaces, with small domestic spaces and larger ritual spaces in close proximity. The areas of Conchopata that can be defined as ritual, using Moore's five-point criterion, are D-shaped temples and large patios. These areas did not contain the highest plant densities or ubiquities. What is perhaps of greater relevance to our discussion is the fact that these areas did contain evidence for the use of plants and the consumption of staple foods. This leads us to the conclusion that these spaces were likely used in ritual events involving food as well as in quotidian moments.

Table 6.7 Table of the top ten combined densities (seeds or fragment/L) by area

Area	Chenopods	Maize cupules	Maize kernels	Parenchyma	*Schinus molle*	Grand total
A174 T1	8.45	0.55	0.35	0.20	7.15	16.70
A180	9.61	0.02	0.10	1.05	2.72	13.50
A176 T1	6.23	3.17	1.23	0.70	1.27	12.60
A186	9.20	1.10	0.45	1.18	0.08	12.00
A40 B	1.21	0.96	1.50	0.79	7.49	11.94
A6	3.48	0.20	0.06	0.42	4.00	8.16
A174	3.90	0.50	2.00	0.30	0.50	7.20
A192	0.31	0.21	1.53	1.14	3.69	6.89
A154	4.22	0.52	0.30	0.08	0.42	5.54
A88	0.52	0.27	0.19	2.56	0.69	4.24

The site areas with the highest densities of plant remains are predominantly the small and enclosed spaces that are proximate to patio groups and open spaces. These areas include architectural spaces EA-174, EA-176, EA-180, and EA-40. Some of these areas also had separate trenches excavated that revealed particularly high concentrations of plants used in brewing *chichas*, such as *molle* and maize. The separated nature of these spaces may have made them ideal locations for the disposal of processed plant material.

Ritual areas that contain more plant remains could be potential brewing areas or ritual food sacrifice areas. The process of brewing should produce several types of botanical signatures: high levels of fuel and brewing plant remains and low abundance of food remains. We did not find this signature in ritual spaces at Conchopata, but that does not mean that brewing did not occur. Rather, the highest densities of plant remains associated with brewing are found in the small domestic areas and "homes," not in the ritual areas, showing that most of the processing was done away from the ritual areas, and presumably brought into these areas at a later time. The brewing process thus appears to have been decentralized, and the amount of recovered botanical evidence is likely not the sum total of all brewing activities at Conchopata, but rather an accumulation of the most recent episodes of brewing.

Feasting Through Botanical Evidence and Ceramics

This chapter's focus on domestic and ritual space also raises a related issue in food consumption, that of feasting. One simple definition of feasting is: "any sharing between two or more people of special food (i.e., foods not generally served at daily meals) in a meal for a special purpose or occasion" (Hayden 2001: 28). This definition leaves open the possibility that a feast can occur in any architectural setting. Thus, the material evidence for feasting behavior may be minimal, or in contrast, it may be overt and significant.

Most arguments about Wari ritualistic feasting and food use have come from evidence of ceramic assemblages rather than direct evidence of botanical remains (such as in Cook and Glowacki 2003). However, the faunal remains from the site do indicate that feasting type consumption of meat did occur in discrete events (Rosenfeld 2012). In contrast, the botanical evidence at Conchopata presents a low density of actual plant remains in its Wari-style ritual center and the ritual spaces tend to be even lower in density relative to the domestic areas.

We find evidence for Wari-style public feasting (making of *chicha de maiz* or *chicha de molle*, large-scale food preparation, and costly discard of meat cuts) at Conchopata because we can satisfy a spatial and a functional requirement needed to make this statement valid. There are patio areas appropriately sized for large feasting activities paired with the presence of small adjacent rooms and "domestic" areas that show evidence of food preparation. Plant species used only as food are found in denser quantities in these domestic rooms than in ritual areas or compounds.

Conclusion

> [Here] I have tried to take one theme—the dynamic between inclusion and exclusion in ritual—and track it through an example drawn from the archaeological record. This allows us to see how communal events could bring people into the group through transformation, and how these could, in turn, be transformed into more hierarchical rituals, which could lead to other powers within the social world. Both of these ritual types create meanings different from those achievable in everyday reality. Thus, even foods consumed during ritual time are different from the same dish consumed in daily life. (Hastorf 2007: 99)

In this chapter, we have analyzed how the use of plants varies across space and how the type of site (urban center, ritual site, or secondary city) can impact the patterning of plant remains. We focused on four important plants *Chenopodium quinoa*, *Schinus molle*, and *Zea mays*, and tuber-based plant foods (parenchyma) that were abundant at Conchopata and give us insight into potential brewing, production, and food preparation practices. *Chichas* were produced at Conchopata; however, the brewing and cooking was likely done in small domestic rooms and spaces close to ritual areas and subsequently brought into these areas for later consumption. Thus, while the large communal spaces may be expected to contain the remains of feast foods and ritual dedications, we tend to find these spaces sparsely covered by plant remains. Rather, the evidence for these events was recovered in discrete rooms to the side of the patios in spaces not overtly marked by fire; the counts of charred wood in ritual areas were similarly low. In order to have this range of results it is necessary to collect blanket samples, as Hastorf (1999) and other paleoethnobotanists have commented, from all spaces—not just those that the excavators deem to be of great significance.

The results from the botanical analysis reveal a mixed picture of consumption and production areas. There are many instances when small-scale ritual activities were likely not visible because they did not produce copious amounts of carbonized materials, did not involve pyro conversion of materials, or were thoroughly cleaned after the fact. We see a slight preference for brewing remains in areas that were likely ritual areas, and a slightly higher preference for food plants in domestic areas. The faunal data provide clearer evidence in this case, with entire ritual deposits of animal remains generally being located outside of domestic areas (Rosenfeld 2012). Plants in certain instances contain less depositional evidence, although as Moore (1989) demonstrated in *chicha* production, it does not have to be the case. Overall the abundance of botanical remains was high in domestic contexts where food and drink were likely made for consumption in both daily and special contexts. In these smaller, more personal spaces, events that trace past behavior may have less overt evidence for ritual practice. However, as noted above, four of the ten samples with the highest densities of botanical remains were labeled as ritual. Thus, contextual definitions illustrate that ritual events occurred in or near domestic space at Conchopata.

This work at Conchopata stands in contrast to paleoethnobotanical research conducted at the Middle Horizon sites of Tiwanaku and Cerro Baúl, where there is greater contextual variation of botanical remains. Thus, this work expands our knowledge of plant use in the past by broadening our understanding of how different site types contain distinct plant remains. The differences between Conchopata and the two other Middle Horizon sites allow us to understand how ritual events varied from context to context and from site type to site type.

At Conchopata, we see that large communal rituals associated with brewing and large amounts of food were more ostensibly visible in the archaeological record. These events presumably emphasized group cohesion and adherence to state norms. Yet, the small seeds left behind in enclosed rooms reveal the intimate events of small-scale rituals. Further research on the multivariate patterning of material remains across different architectural spaces will provide additional insights into how ritual and domestic life blended into one another to provide people with the means to commemorate small and large occasions in their lives. In the end, the remains recovered from Conchopata leave us with the impression that domestic and ritual life continuously blended together to leave behind a mixed record of daily activity in an early urban center.

Acknowledgements The authors would like to thank William Isbell, Anita Cook, and the Conchopata excavation team for their work on collecting and housing the authors while the samples were processed in Peru. Christine Hastorf was instrumental in giving us this opportunity all these years ago, and the team of undergraduates that helped with some of the sorting of materials. Much of the sorting was done in Ripon by Kelsey Green under the supervision of Whitehead. The Field Museum of Chicago, donated several hours of SEM time, and the Dean of Faculty at Ripon College who donated laboratory and research funds to complete the final analysis. This manuscript was reviewed and commented on by an anonymous reviewer and Maria Bruno.

References

Aufderheide, A. C., Buikstra, J. E., Cartmell, L., & Weems, C. (1991). *The Prehistory of Andean Coca Leaf Chewing Practices.* Paper presented at the 90th annual meeting of the American Anthropological Association, Chicago, November 20–24.

Bragayrac D. E. (1991). Archaeological Excavations at Vegachayoq Moqo Sector of Wari. In W. H. I. Isbell & G. McEwan (Eds.), *Wari Administrative Structure: Prehistoric Monumental Architecture and State Government* (pp. 71–80). Washington D.C.: Dumbarton Oaks Research Library and Collection.

Browman, D. L. (1976). Demographic correlations of the Wari conquest of Junin. *American Antiquity, 41*(4), 465–477.

Bruno, M. C. (2006). A morphological approach to documenting the domestication of *Chenopodium* in the Andes. In M. Zeder, D. Bradley, E. Emshwiller, & B. Smith (Eds.), *Documenting domestication: New genetic and archaeological paradigms* (pp. 32–45). Berkeley: University of California Press.

Bruno, M. C., & Ramos, M. (2009). *Plant Remains from Residential and Mortuary Contexts of Mollo Kontu, Tiwanaku.* Paper presented at the Society for American Archaeology Annual Meeting, Atlanta, Georgia, April 23.

Bruno, M. C., & Whitehead, W. T. (2003). *Chenopodium* cultivation and formative periods agriculture at Chiripa, Bolivia. *Latin American Antiquity, 14*(3), 339–355.

Cartmell, L. W., Auderheide, A. C., Springfield, A., Weems, C., & Arriaza, B. (1991). The frequency and antiquity of prehistoric coca leaf chewing practices in northern Chile: Radioimmunoassay of a cocaine metabolite in human-mummy hair. *Latin American Antiquity, 2,* 260–268.

Chiou, K. L., Hastorf, C. A., Bonavia, D., & Dillehay, T. D. (2014). Documenting cultural selection pressure changes on chile pepper (*Capsicum baccatum* L.) seed size through time in coastal Peru (7,600 B.P.–Present). *Economic Botany, 68*(2), 190–202.

Cook, A. (2001). Wari D-shaped structures, sacrificial offerings, and divine rulership. In E. Benson & A. Cook (Eds.), *Ritual sacrifice in ancient Peru* (pp. 119–163). Austin: University of Texas Press.

Cook, A., & Benco, N. (2000). Vasijas para la fiesta y la fama: producción artesanal en un centro urbano Huari. In P. Kaulicke, & W. Isbell (Eds.), *Boletín de arqueología PUCP, No 4. Wari y Tiwanaku: Modelo vs. evidencias. Primera parte* (pp. 489–504). Lima: Pontificia Universidad Católica del Perú.

Cook, A., & Glowacki, M. (2003). Pot, politics, and power: Wari ceramic assemblages and imperial administration. In T. L. Bray (Ed.), *The archeology and politics of food and feasting in early states and empires* (pp. 173–202). New York: Kluwer Academic/Plenum.

Cook, A., & Isbell, W. H. (2000). Emergency Excavations at Conchopata: Wari Temple Architecture and Iconography in Ayacucho, Peru: Project Grant Report, 1999–2000. http://www.doaks.org/PCProjectGrants/cook/cook.html8, July 2016.

Dean, E., & Kojan, D. (2001). Ceremonial households and domestic temples: "Fuzzy" definitions in the Andean Formative. *Kroeber Anthropological Society Papers, 85,* 109–135.

Earle, T., & Jennings, J. (2013). Remodeling Wari Political Economy. In *Boletín de arqueología PUCP, No. 16. Los rostros de Wari: Perspectivas interregionales sobre el Horizonte Medio* (pp. 209–226). Lima: Pontificia Universidad Católica del Perú.

Finucane, B., Agurto, P. M., & Isbell, W. (2006). Human and animal diet at Conchopata, Peru: Stable isotope evidence for maize agriculture and animal management practices during the Middle Horizon. *Journal of Archaeological Science, 33,* 1766–1776.

Glowacki, M. (2002). The Huaro archaeological site complex: Rethinking the Wari occupation of Cuzco. In W. H. Isbell & H. Silverman (Eds.), *Andean archaeology I: Variations in sociopolitical organization* (pp. 267–286). New York: Kluwer Academic/Plenum.

Goldstein, D. J., & Coleman, R. C. (2004). *Schinus molle* L. (Anacardiaceae) *Chicha* production in the Central Andes. *Economic Botany, 58*(4), 523–529.

Green, K. R., & Whitehead, W. T. (2006). *Domesticated Plant Remains from Conchopata, a Wari Ceremonial Site in the Ayacucho Valley*. Poster Session presented at the 71st Annual Meeting of the Society for American Archaeology, San Juan, Puerto Rico.
Hastorf, C. A. (1990). The effect of the Inka State on Sausa agricultural production and crop consumption. *American Antiquity, 55*, 262–290.
Hastorf, C. A. (1999). Recent research in paleoethnobotany. *Journal of Archaeological Research, 7*(1), 55–103.
Hastorf, C. A. (2001). Studying ritual in the past. In past ritual and the everyday. *Kroeber Anthropological Society Papers, 85*, 1–15.
Hastorf, C. A. (2007). Archaeological Andean rituals: Performace, liturgy, and meaning. In E. Kryriakidis (Ed.), *The archaeology of ritual* (pp. 77–107). Los Angeles, CA: UCLA Cotsen Press.
Hastorf, C. A., Whitehead, W. T., Bruno, M. C., & Wright, M. (2006). The movements of maize: The path of maize into Middle Horizon Tiwanaku, Bolivia. In J. Staller, R. Tykot, & B. Benz (Eds.), *Histories of maize: Multidisciplinary approaches to the prehistory, linguistics, biogeography, domestication, and evolution of maize* (pp. 429–447). San Diego, CA: Academic Press/Elsevier Press.
Hawkes, C. (1954). Archaeological theory and method: Some suggestions from the Old World. *American Anthropology, 56*(2), 153–168.
Hayden, B. (2001). Fabulous feast: A prolegomenon to the importance of feasting. In M. Dietler & B. Hayden (Eds.), *Feasts: Archaeological and ethnographic perspectives on food, politics, and power* (pp. 23–64). Washington: Smithsonian Institutional Press.
Isbell, W. H. (2004). Mortuary preference: A Wari culture case study from Middle Horizon Peru. *American Antiquity, 15*(1), 3–30.
Isbell, W. H. (2008). A community of potters, or multi-crafting wives of polygynous lords. In I. Shimada (Ed.), *Craft production in complex societies: Multicraft and producer perspectives (Foundations of Archaeological Inquiry)* (pp. 68–96). Salt Lake City: University of Utah Press.
Isbell, W. H. (2009). Wari: A new direction in central Andean urban evolution. In L. R. Manzanilla & C. Chapdelaine (Eds.), *Domestic life in prehispanic capitals: A study of specialization, hierarchy, and ethnicity* (pp. 197–219). Ann Arbor: University of Michigan, Museum of Anthropology.
Isbell, W. H., & Cook, A. G. (2002). A new perspective on Conchopata and the Andean Middle Horizon. In H. Silverman & W. H. Isbell (Eds.), *Andean archaeology II: Art, landscape, and society* (pp. 249–305). New York: Kluwer Academic/Plenum.
Isbell, W. H., & McEwan, G. F. (Eds.). (1991). *Wari administrative structure: Prehistoric monumental architecture and state government*. Washington, D.C.: Dumbarton Oaks Research Library and Collection.
Isbell, W. H., & Schreiber, K. J. (1978). Was Wari a State? *American Antiquity, 43*(3), 372–389.
Jennings, J., & Bowser, B. (Eds.). (2008). *Drink, power, and society in the Andes*. Gainseville: University Press of Florida.
Jennings, J., & Craig, N. (2001). Politywide analysis and imperial political economy: The relationship between valley political complexity and administrative centers in the Wari empire of the central Andes. *Journal of Anthropological Archaeology, 20*, 479–502.
Kaulicke, P., & Isbell, W. H. (Eds.). (2000). *Wari y Tiwanaku: Modelo vs. evidencias. Primera parte*: Boletín de arqueología PUCP, No 4. Lima.
Kaulicke, P., & Isbell, W. H. (Eds.). (2001). *Wari y Tiwanaku: Modelo vs. evidencias. Segunda parte*: Boletín de arqueología PUCP, No 5. Lima.
Kintigh, K. W., & Altschul, J. H. (2010). Sustaining the digital archaeological record. *Heritage Management, 3*(2), 264–274.
Knobloch, P. J. (2000). Wari ritual power at Conchopata: An interpretation of *Anadenanthera colubrina* iconography. *Latin American Antiquity, 11*(4), 387–402.
Knudson, K. J., & Buikstra, J. E. (2007). Residential mobility and resource use in the Chiribaya polity of Southern Peru: Strontium isotope analysis of archaeological tooth enamel and bone. *International Journal of Osteoarchaeology, 17*(6), 563–580.

Kolata, A. L. (1993). *The Tiwanaku: Portrait of an Andean civilization*. Cambridge: Blackwell.
Lanning, E. (1967). *Peru before the Incas*. Englewood Cliffs, New Jersey: Prentice-Hall.
Leoni, J. B. (2005). La veneración de montañas en los Andes preincaicos: el caso de Ñawinpukyo (Ayacucho, Perú) en el período Intermedio Temprano. *Chungara Revista de Antropología Chilena, 3*(2), 151–164.
Leoni, J. B. (2006). Ritual and society in early Intermediate Period Ayacucho: A view from the site of Nawinpukyo. In W. H. Isbell & H. Silverman (Eds.), *Andean archaeology* (Vol. III, pp. 279–306). New York: Springer.
Lumbreras, L. G. (1974). *The peoples and cultures of ancient Peru* (B. J. Megers, Trans.). Washington, D.C.: Smithsonian Institution Press.
Machaca, G. (1997). *Secuencia cultural y nuevas evidencias de formación urbana en Ñawinpuquio*. Tesis de Licenciatura. Universidad de San Cristobal de Huamanga, Ayacucho, Peru.
McEwan, G. F. (1990). Some formal correspondences between the imperial architecture of the Wari and Chimu cultures of ancient Peru. *Latin American Antiquity, 1*(2), 97–116.
McEwan, G. F. (1996). Archaeological investigations at Pikillacta, a Wari site in Peru. *Journal of Field Archaeology, 23*(2), 169–186.
McEwan, G. F. (1998). The function of Niched Halls in Wari architecture. *Latin American Antiquity, 9*(1), 68–86.
Menzel, D. (1964). Style and time in the Middle Horizon. *Ñawpa Pacha: Journal of Andean Archaeology, 2*(1), 1–106.
Moore, J. (1989). Pre-Hispanic beer in coastal Peru: Technology and social context of prehistoric production. *American Anthropologist, 91*(3), 682–695.
Moore, J. (1996). *Architecture and power in the Prehispanic Andes: The archaeology of public buildings*. Cambridge: Cambridge University Press.
Morehart, C. T., & Morell-Hart, S. (2013). Beyond the ecofact: Toward a social paleoethnobotany in Mesoamerica. *Journal of Anthropological Archaeology, 22*(2), 483–511.
Moseley, M. E. (1992). *The Incas and their ancestors: The archaeology of Peru*. New York: Thames and Hudson.
Moseley, M. E., Nash, D. J., Williams, P. R., deFrance, S. D., Miranda, A., & Ruales, M. (2005). Burning down the brewery: Establishing and evacuating an ancient imperial colony at Cerro Baúl, Peru. *Proceedings of the National Academy of Sciences, 102*(48), 17264–17271.
Ochatoma Paravicino, J., & Cabrera Romero, M. (2000). Arquitectura y areas de actividad en Conchopata. In P. Kaulicke & W. Isbell (Eds.), *Boletín de arqueología PUCP, No 4. Wari y Tiwanaku: Modelo vs. evidencias. Primera parte* (pp. 449–488). Lima: Pontificia Universidad Católica del Perú.
Ochatoma Paravicino, J., & Cabrera Romero, M. (2002). Religious ideology and military organization in the iconography of a D-shaped ceremonial precinct at Conchopata. In H. Silverman & W. H. Isbell (Eds.), *Andean archaeology II: Art, landscape, and society* (pp. 225–247). New York: Kluwer Academic/Plenum.
Parrish, N. (1999). Field Report on Floatation and Preliminary Paleoethnobotanical Analysis from Conchopata 1999. Report on file. New York: William Isbell, Binghamton University.
Pearsall, D. M. (2000). *Paleoethnobotany: A handbook of procedures* (2nd ed.). San Diego, California: Academic Press.
Plowman, T. (1984). The origin, evolution and diffusion of Coca, *Erythroxylum* spp., in South and Central America. In D. Stone (Ed.), *Pre-Columbian Plant Migration* (pp. 125–163, Papers of the Peabody Museum of Archaeology and Ethnology 76). Cambridge: Harvard University.
Plowman, T. (1985). Coca chewing and the botanical origins of coca (*Erythroxylum* spp.) in South America. In D. Paccini & C. Franquemont (Eds.), *Coca and cocaine: Effects of people and policy in Latin America* (pp. 5–34, Cultural Survival Report 23). Cambridge: Cultural Survival.

Popper, V. (1989). Selecting quantitative measurements. In C. A. Hastorf & V. Popper (Eds.), *Current paleoethnobotany: Analytical methods and cultural interpretations of archaeological plant remains* (pp. 1–16). Chicago, IL: University of Chicago Press.

Pozzi-Escot, D. (1991). Conchopata: A community of potters. In W. H. Isbell & G. McEwan (Eds.), *Wari administrative structure: Prehistoric monumental architecture and state government* (pp. 81–92). Washington, D.C.: Dumbarton Oaks Research Library and Collection.

Renfrew, J. (1973). *Paleoethnobotany: The prehistoric food plants of the Near East and Europe*. New York: Columbia University Press.

Rosenfeld, S. (2012). Wealth display and local power in the Wari Empire. *Ñawpa Pacha: Journal of Andean Archaeology, 32*(1), 131–164.

Rowe, J. H., Collier, D., & Willey, G. R. (1950). Reconnaissance notes on the site of Wari, near Ayacuchu, Peru. *American Antiquity, 16*(2), 120–137.

Sayre, M. (2014). Ceremonial plants in the Andean region. In A. Chevalier, E. Marinova, & L. Peña-Chocarro (Eds.), *Crops and people: Choices and diversity through time* (pp. 369–373). Brussells: European Science Foundation/Early Agricultural Remnants and Technical Heritage Program.

Sayre, M., Goldstein, D., Whitehead, W. T., & Williams, P. R. (2012). A marked preference: Chicha de Molle and Wari State Consumption Practices. *Ñawpa Pacha: Journal of Andean Archaeology, 32*(2), 231–258.

Sayre, M., & Whitehead, W. T. (2002). *New Paleoethnobotanical Evidence from Conchopata: A Wari Site.* Paper presented at the 68th Annual Meeting of the Society for American Archaeology, Milwaukee, Wisconsin.

Schreiber, K. J. (1992). *Wari imperialism in Middle Horizon Peru* (Anthropological Papers, No. 87). Ann Arbor: Museum of Anthropology, University of Michigan.

Schultes, R., & Hoffman, R. (2001). *Plants of the Gods: Their sacred, healing, and hallucinogenic powers* (2nd ed.). New York: Healing Arts Press.

Simpson, B. B., & Conner-Ogorzaly, M. (2013). *Plants in our World: Economic botany* (4th ed.). New York: McGraw-Hill Education.

Spielmann, K. A., & Kintigh, K. W. (2011). The digital archaeological record: The potentials of archaeozoological data integration through tDAR. *SAA Archaeological Record, 11*(1), 22–25.

Tung, T. A. (2007). Trauma and violence in the Wari empire of the Peruvian Andes: Warfare, raids, and ritual fights. *American Journal of Physical Anthropology, 133*(3), 941–956.

Tung, T. A. (2008). Dismembering bodies for display: A bioarchaeological study of trophy heads from the Wari site of Conchopata, Peru. *American Journal of Physical Anthropology, 136*(3), 294–308.

Tung, T. A., & Cook, A. (2006). Intermediate-elite agency in the Wari empire: The bioarchaeology and mortuary evidence. In C. M. Elson & R. A. Covey (Eds.), *Intermediate elites in the Pre-Columbian states and empires* (pp. 68–93). Tucson: The University of Arizona Press.

Van Buren, M. (1992). Wari administrative structure: Prehistoric monumental architecture and state government. *American Anthropologist, New Series, 94*(3), 756–757.

VanDerwarker, A. M., & Peres, T. M. (Eds.). (2010). *Integrating zooarchaeology and paleoethnobotany: A consideration of issues, methods, and cases*. New York: Springer.

Wilbert, J. (1993). *Tobacco and shamanism in South America*. New Haven, CT: Yale University Press.

Williams, P. R. (2001). Cerro Baúl: A Wari center on the Tiwanaku frontier. *Latin American Antiquity, 12*(1), 67–83.

Williams, P. R. (2002). Rethinking disaster-induced collapse in the demise of the Andean highland states: Wari and Tiwanaku. *World Archaeology, 33*(3), 361–374.

Williams, P. R., & Nash, D. J. (2002). Imperial interaction in the Andes: Wari and Tiwanaku at Cerro Baúl. In W. H. Isbell & H. Silverman (Eds.), *Andean archaeology I: Variations in sociopolitical organization* (pp. 243–266). New York: Kluwer Academic/Plenum.

Wright, M., Hastorf, C. A., & Lendstrom, S. (2003). Pre-Hispanic agriculture and plant use at Tiwanaku: Social and political implications. In A. L. Kolata (Ed.), *Tiwanaku and its Hinterlands* (pp. 384–403). Washington D.C.: Smithsonian Institution Press.

Yarnell, R. A. (1970). Paleo-ethnobotany in America. In D. Brothwell & E. Higgs (Eds.), *Science in archaeology: A survey of progress and research* (2nd ed., pp. 215–228). New York: Praeger.

Chapter 7
Ritual Time: The Struggle to Pinpoint the Temporality of Ritual Practice Using Archaeobotanical Data

Christopher T. Morehart

Ritual is about time. Rituals are carefully choreographed and highly structured performances. Understanding ritual requires a consideration of its spatiotemporal characteristics. For decades, a functionalist approach dominated the study of ritual in anthropology and archaeology (Fowles 2010). Echoing a legacy in the social and historical sciences, researchers searched for the societal functions behind ritual and other forms of symbolic behavior and conceptual models, especially in enhancing ecological interaction or in fostering social solidarity (i.e., Douglas 1996; Durkheim 1915; Harris 1966; Rappaport 1968; Turner 1969). Influenced by these approaches, many works stress a macroscalar, society-wide interpretive lens to position ritual in reconstructions of the past. However, the conceptual frameworks that guided ritual behavior were often considered epiphenomenal, and the detailed characteristics of rituals' material assemblages were commonly subordinated in macroscalar narratives of social organization and developmental change.

A tension thus exists in understanding past ritual in anthropological and archaeological literature. In a crude sense a dichotomy between more *etic* versus *emic* approaches to accessing the past can be painted. Derived from the words phonemic and phonetic, emic and etic capture the extent to which statements of cultural phenomena are meaningfully legible by differentially situated viewers, as the voiced bilabial stop /b/ would be phonetically contrasted from the voiced bilabial fricative /v/, though both sounds would be recognized as phonemically the same sound by a Spanish speaker (as allophones). Hence, Harris (1979: 32) expanded and popularized this distinction to characterize insider versus outsider views of culture. He stressed that the role of the anthropologist is to construct etic models meaningful to scientists, though perhaps less meaningful to a member of the cultural group. From a similar perspective, tuned specifically to the interplay between cultural practices and beliefs (i.e., religion and ritual) and ecology,

C.T. Morehart (✉)
Department of Anthropology, Arizona State University, Tempe, AZ, USA
e-mail: Christopher.Morehart@asu.edu

M.P. Sayre and M.C. Bruno (eds.), *Social Perspectives on Ancient Lives from Paleoethnobotanical Data*, DOI 10.1007/978-3-319-52849-6_7

Rappaport (1979) argued that "cognized models" represented the world as seen by a member of a cultural group, whereas an "operationalized model" is one constructed by an anthropologist for analytical and comparative purposes. Despite Rappaport's and Harris' (albeit distinctive) materialisms, this was a wide view of anthropology's role as a science. Lévi-Strauss, for example, proposed "to show, not how men think in myths, but how myths operate in men's minds without them being aware of the fact" (Lévi-Strauss 1975: 11–12). Indeed, Harris's application of emic and etic is a co-option of Kenneth Pike's concepts, the original sense of which was more akin to Lévi-Strauss's structural analyses (D'Andrade 1995: 19). Nevertheless, the distinction underlies epistemological differences and disparate commitments to the very ontological status of culture itself for many scholars, which at times has driven a wedge into the fabric of anthropological scholarship and intellectual identity (see Roscoe 1995).

Archaeology historically has maintained a similar, though different, debate, conditioned by the inability of modern-day researchers to get into the heads of past people or to document seemingly intangible domains of culture. Conjuring the specter of Hawke's ladder or the ambiguous "rest of culture" surrounding Steward's core (see Robb 1998), these aspects of past society are not just considered epiphenomena. They are viewed as empirically inaccessible. This perspective is very obvious in paleoethnobotany, the study of past peoples' interactions with the plant world. That is, examining ritual practices, both beliefs and behaviors, is drastically underrepresented in paleoethnobotany (see Morehart and Morell-Hart 2015). Exceptions exist, of course. Hastorf and Johannessen (1991), for example, demonstrate how wood selection in the ancient Andes passed through cultural, folk taxonomic filters and also how particular taxa held unique symbolic values. Morehart et al. (2005) examined the ritual use of pine by the Lowland Maya and suggested that its burning was not only pervasive but had little utilitarian function. (McNeil 2006), working in the Maya city of Copan, and Montúfar Lopez (2006), working in the Mexica city of Tenochtitlan, compared plant remains from ritual contexts with ethnographic and iconographic information to reconstruct the types of rituals and the cosmological significance of specific taxa.

Most paleoethnobotanical research, however, historically and globally focuses on the origin and nature of subsistence economies and environmental adaptation. The exclusive utilitarian attention can squeeze culture out of our understandings of how people in the past used plants. Furthermore, a potentially Eurocentric and modern view of time results from this adaptation-centric perspective. Optimized economic behaviors are seen as temporally accessible and can be easily modeled using conceptual apparatuses of the present (see Lucas 2005). Ritual behaviors, however, occupy qualitatively different temporal and spatial realities—governed by social and religious meanings seemingly inaccessible from archaeological scrutiny. Ignoring ritual not only omits significant aspects of the past from our narratives. It in fact requires archaeologists to overlook empirical data from a highly physical form of behavior. That a conceptual construction as abstract as the "state" commonly is viewed as more accessible to archaeologists than cultural beliefs materialized very concretely in a structured deposit, such as a burial, is both unfortunate and remarkable.

This chapter is an exercise in understanding ritual's temporality from a paleoethnobotanical perspective. I suggest that a consideration of when a rite occurred is a significant aspect of understanding ritual as a practice and, hence, is a prerequisite to any subsequent treatment of ritual's micro- and macrosocial roles. I examine two aspects of temporality at least partly using archaeobotanical remains, which I refer to as seasonality and symbolism. Seasonality denotes the specific time during an annual cycle in which a particular behavior occurred. Symbolism simply recognizes that different natural resources, individually and together, are allotted rich and sometimes multiple meanings by people—meanings that are commonly connected to spiritual entities in spatially- and temporally ordered cosmologies. I explore these two approaches not as a means to contrast etic or operationalized models from emic or cognized ones. Indeed, these dimensions of ritual not surprisingly overlap, and understanding the temporality of ritual requires both these components. Rather, this paper offers the embryo of an approach: analytical and methodological, with, hopefully, substantive results. This exercise reveals the intrinsic challenges in pinpointing ritual time not only using archaeobotanical data but archaeological data more generally. The challenges themselves, however, are interpretively transformative. Rather than revealing the impossibility of understanding ritual and accessing the immaterial, this paper offers an opportunity to reflect on ways to access qualitatively different temporalities and spatialities.

An Epiclassic Period Shrine

I examine these dimensions of temporality by considering an Epiclassic period shrine in the northern Basin of Mexico (Fig. 7.1). The Epiclassic period in central Mexico dates to around AD 600–900 and marks the period between the decline of the powerful Teotihuacan state and the rise of the complex mosaic of Postclassic period city-states that would eventually lead to the formation of the Aztec empire (Diehl and Berlo 1989). It is often considered a time of instability, immigration, conflict, and change. In the northern Basin of Mexico, regional population densities declined and new centers were common in defensible hilltop locations, which might indicate regional violence (Gorenflo and Sanders 2007; Parsons et al. 2008; Sanders et al. 1979).

My work on Epiclassic landscapes began unexpectedly during a project on a much later Middle Postclassic (AD 1200–1400) raised field farming system in Lake Xaltocan (Morehart 2010). During survey in 2007, we stumbled upon a heavily looted site, which we designated Non-Grid 4 (Fig. 7.1). Large, eroded sherds of bowls and incense burners littered the surface, as did figurines and human remains, especially fragments of cranial bones. Without these looters pits we may never had stumbled across it. With the exception of a slight rise in the center of the area, the site was completely buried. This concentration of materials was just north of a

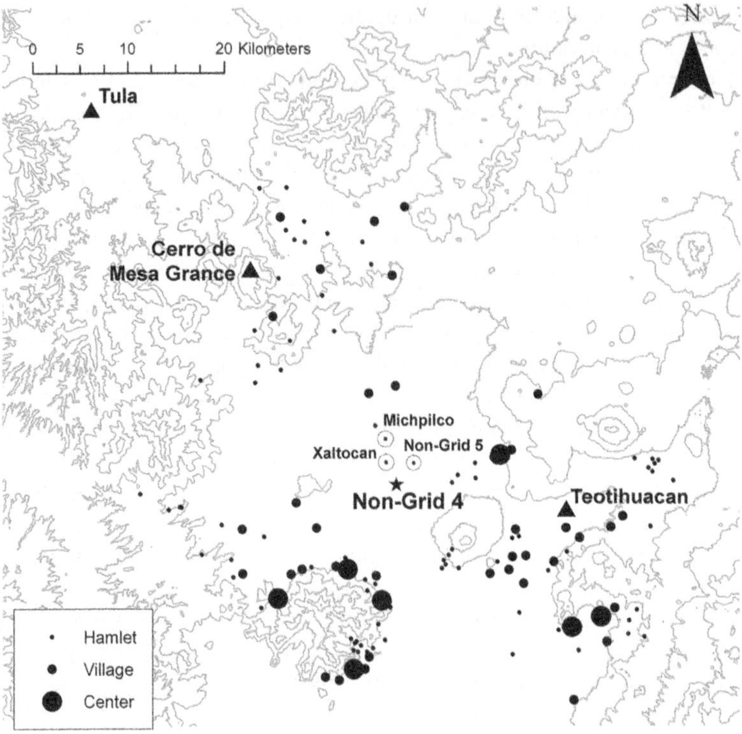

Fig. 7.1 Map of Epiclassic Sites in the Northern Basin of Mexico. Michpilco and Non-Grid 5 are small sites with Epiclassic occupations. Xaltocan is principally a Postclassic site, though it likely overlays an earlier component

series of freshwater springs that still existed up to the 1950s and was situated along what would eventually be a major canal for the later raised field system.

Our excavations revealed a large, amorphous platform created by building up a solid mass of limestone fill, not unlike some of the nearby hamlets (Morehart et al. 2012). Along the edges of the platform, at the bottom of the former lake, we documented smashed vessels and incense burners as well as random postcranial bones. These vessels, distinctive types dating to the Epiclassic period and its transition into the Early Postclassic, dominated all excavation levels as well as the looted surface. No evidence was found to indicate that the area was used for domestic purposes. With the exception of some limited Postclassic and even Colonial period pottery recovered from the ground surface, this area was used most intensively during the Epiclassic period. The earliest AMS date, from pine charcoal associated with a smashed brazier at the bottom of the platform fill, dates the shrine's construction between AD 550 and 660 (Morehart et al. 2012: 437).

Although most of our excavations uncovered just sporadic and disarticulated human remains, in one excavation unit, E10N4, we discovered buried lines of skulls (Morehart et al. 2012). Under each cranium we found one or two neck vertebra, though no other bones from the body. Each skull either faces east or seems to have shifted from an eastern orientation following deposition. Directly under the first burial we uncovered two additional crania that also face east. The crania are next to a large, broken standing brazier. Over the face of one skull, ritual practitioners placed the face of a figurine, likely a deity effigy from an incense burner (Fig. 7.2). AMS dates place the burials between AD 660 and 890, in the Epiclassic period (Morehart et al. 2012: 437). Overall, at least 31 individuals are represented in the

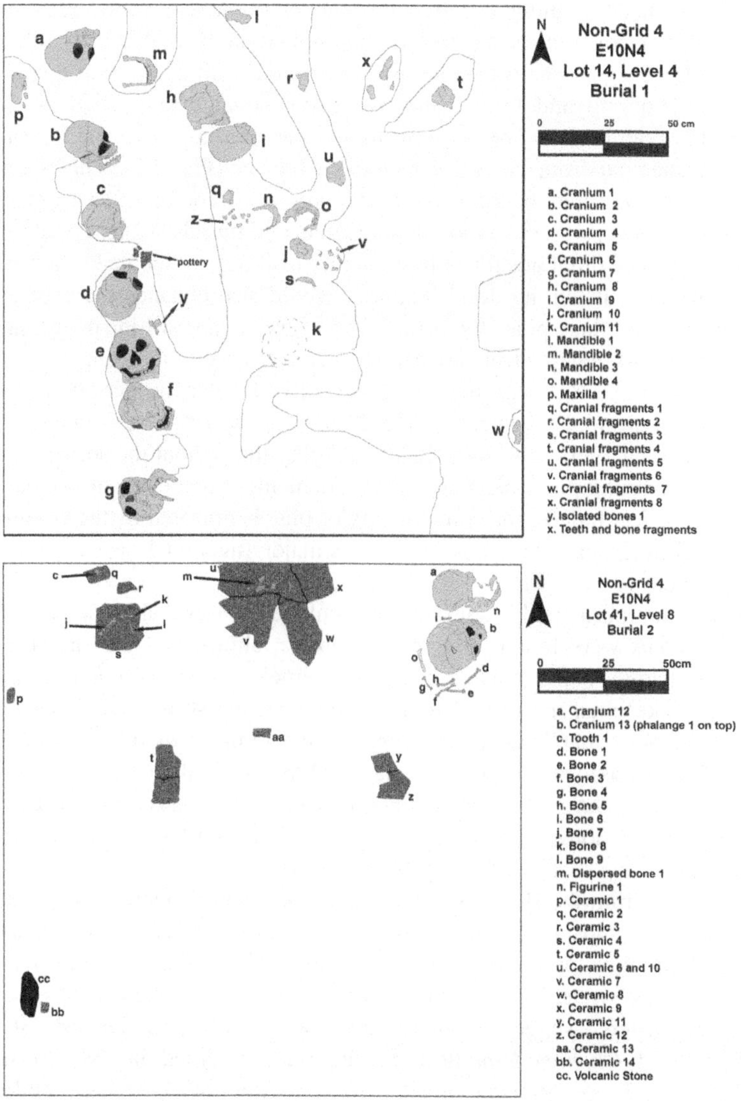

Fig. 7.2 Burial deposits uncovered in E10N4 during 2007–2008 season

skeletal assemblage, and almost all are male adults. However, we returned to the site in 2012 for more intensive, horizontal excavations. Analysis of these data is ongoing, but well over 100 additional individuals were recovered.

Archaeobotanical Remains

Archaeobotanical sampling occurred throughout our excavations. Fifty-two soil samples were recovered for macrofloral and pollen analysis. Virtually all samples (96%) contained pine charcoal. Pine was recovered from excavation levels within the construction fill and from distinctive depositional events, such as smashed braziers and both burials. The soil next to the lower burial, for example, was black with pine charcoal from the adjacent broken brazier (Fig. 7.3). The presence of some pine charcoal may be the result of the process of cooking down limestone to use as fill, but its direct association with braziers and burials suggests that pine also was an element of ritual practice. Pine is highly resinous, which makes it perfect for kindling incense. Its strong floral fragrance would also enhance the experience of ritual itself. Burning pine for ritual was very common in many areas of Mesoamerica (Morehart et al. 2005). The Otomi, an ethnic group said to have settled in the Lake Xaltocan region by the Early Postclassic period, worshipped *Otonteuctli*, a deity of warfare, fire and death, who was also associated with pine (Carrasco Pizana 1950: 138–140). Pine's ubiquity throughout the shrine's use may suggest that burning pine charcoal was a fundamental ritual element—a temporally invariant practice. Indeed, the commonality of pine is mirrored by the commonality of sherds from incense burners, which is a major functional category of pottery from the site.

Other taxa, particularly domesticated cultigens, were also fairly common. Although beans were found only in one sample, almost half the samples (40%) contained charred maize remains, including kernels, individual cupules, and cobs (Fig. 7.4). Like pine, maize was found throughout the shrine, including with the burials. I suggest that these items represent the remains of offerings of food to spiritual entities associated with this place (Morehart et al. 2012). The seeds of a species of *Chenopodium* were also found. However, they lacked the morphological attributes of the domesticated *Chenopodium* (*Huauzontle*) and could represent naturally occurring local wild flora (Minnis 1981).

Pollen data from the shrine also appear to document deliberate ritual activities. Not surprisingly, most of the shrine pollen documents regional vegetation and local wetland flora. Pine pollen dominates, and the pollen assemblage also includes grasses, sedges, water lilies, and cattails. Cheno-Am pollen grains occur in aggregate clusters, suggesting they are the remnants of anthers of either *Amaranthus* or *Chenopodium* flowers, and were recorded in 70% of samples ($n = 14$) (Fig. 7.5a). An even greater concentration of Asteraceae pollen was identified to the genus *Tagetes*, most likely *T. erecta*, also known as *cempoalxochitl* or Aztec Marigold (Fig. 7.5b). *Tagetes* sp. pollen occurred in 80% of samples

Fig. 7.3 Soil blackened with pine charcoal and ash from brazier surrounding two crania designated Burial 2. Note effigy face over north-most cranium (compare with Fig. 7.2, bottom)

Fig. 7.4 Carbonized maize (*Zea mays*) cob from shrine, marks indicate 1 mm

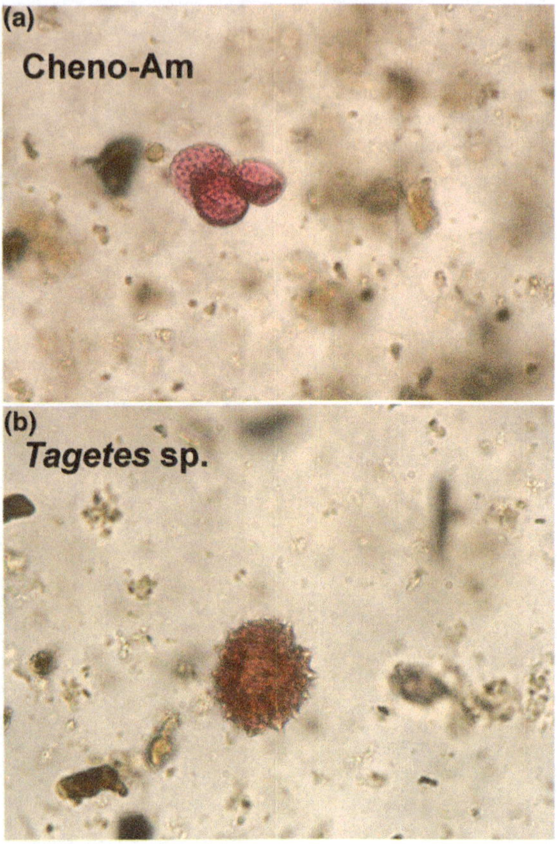

Fig. 7.5 Pollen from probably floral offerings. **a** Cheno-Am pollen aggregate. **b** *Tagetes* cf. *erecta*

(n = 16). In the unit with the intact crania, E10N4, *Tagetes* sp. was documented in all levels except the shallowest and the deepest. However, the concentration of *Tagetes* is highest in levels associated with the crania burials compared to surrounding strata (see Morehart et al. 2012: 442).

Seasonality and Symbolism

To understand the temporality of these ritual practices requires we consider both seasonality and symbolism. Examining seasonality seems to be the most accessible line of reasoning. Relating archaeobotanical remains to annual climatological shifts has long offered a means to model variation in human–environmental interaction. The presence of agricultural products, such as maize and beans, might indicate their deposition not long after harvest, perhaps as offerings of repayment to deities, a common practice throughout Mesoamerica. Of course, deciding on appropriate ecological and ethnographic analogues is not simple in an ecologically diverse region where various types of crops with different cycles were cultivated.

Despite variation, most maize in central, highland Mexico is either a short or an early maturing variety (Charlton 1970; Sanders et al. 1979: 233; Wellhausen et al. 1952). Although perhaps counterintuitive, shorter maturation typically does not relate to multiple harvests. Instead, these types, which mature in about 3 months, are suited to areas where irrigation is not practical. These early maturing types are low yielding, but faster maturation reduces the length of time the crop is susceptible to unpredictable shifts in rainfall. These types are planted beginning in April to as early as June and are harvested from mid to late summer.

The longer maturing variety, often called *chalqueño*, is cultivated in irrigated areas and places with high soil humidity. *Chalqueño* maize has a high yield and is planted in late spring and harvested in 6 months (Sanders et al. 1979: 233; Wellhausen et al. 1952). Around the turn of the century, Santamaría (1912) observed that chinampa farmers in the southern Basin of Mexico grew long maturing varieties that they harvested in November together with beans. Sanders et al. (1979: 235) suggest that this type of maize was in existence archaeologically by the Epiclassic period. Morphologically, the maize from the shrine differs little from measurements of *chalqueño* recorded by Benz (1986: 353). However, the remains show overlap with many kinds of highland Mexican maize, especially in row number (ranging between 14 and 20). Row number would not change due to firing, whereas other features may shrink or become distorted (Benz 1994). However, given the wetland environment of the shrine, it is quite possible that local inhabitants cultivated a similarly adapted maize. Indeed, maize agriculture in the immediate area today is heavily irrigated and employs a longer maturing variety that is planted in May and harvested between August and September. Thus, these comparative data might suggest that rituals involving maize offerings occurred in the final months of the rainy season during the late summer and early fall. In the seventeenth century, Ruiz de Alarcón (1984: 53), for example, wrote that "the time when there is most manifestly formal idolatry is at the end of the rainy season".

The seasonality of flower taxa represented in the shrine by pollen supports this provisional conclusion. *Cempoalxochitl*, also known today as *Flor de Muerto*, is harvested today en masse in late October for annual Day of the Dead festivities in Mexico. The indigenous solar calendar offers a different timing for similar practices during the ninth month Miccailhuitl, which was celebrated with feasts in honor of the dead (Keber 1995: 141–143). The sixteenth century Spanish priest Sahagún recorded this month as Tlaxochimaco, or "The Offering of Flowers." The feasts during this month involved the collection of every kind of flower in bloom, which were strung together in ornaments and used to adorn the warrior God Uitzilopochtli (Sahagún 1951: 103). Sahagún states that generally flowers are used soon after they blossom: "At this time they are required, desired, coveted, needed" (Sahagún 1963: 214). Soon after their use "they fade, shed petals, drop petals, darken, wither" (Sahagún 1963: 214).

Information recorded by Sahagún and others (i.e., Durán 1971) reveal that the timing of rituals cannot be understood only in terms of twentieth century ecological analogies. In addition to ecological constraints on available resources, ritual behaviors were carefully choreographed according to both solar and ceremonial calendars. Examining such ritual symbolism, however, forces us to assume long-term continuity from the Epiclassic period to the Early Colonial period in central Mexico. When integrated with other archaeological data from the shrine, such assumptions are not completely farfetched. However, the goal of this exercise is not to establish beyond a doubt the exact meaning of past rituals or even long-term continuity. Rather, integrating comparative information on the symbolism of ritual offerings suggests different temporal patterns in the nature of ritual practices than if we only relied upon data pointing to ecological cycles.

Several effigy figurines of deities, most notably Tlaloc, God of water and rain, were recovered during work at the shrine (Fig. 7.6). These figurines differ in detail. For example, one is in the form of a bird, likely a heron, an animal associated with Tlaloc (Sahagún 1951: 3). Nevertheless, all have the distinctive ringed eyes and other facial features of Tlaloc. The effigy face placed over the cranium in one of the burials lacks attributes of Tlaloc but appears to share some features in common with the deity Xipe Totec, the God of fertility and rebirth, especially its inset eyes and double lips. In other representations of Xipe, these elements indicate the wearing of flayed skin.

Gods related to both Tlaloc and Xipe Totec appeared in the cosmological repertoire for both earlier and later peoples in central Mexico (Arnold 1999; Broda 1971; Durán 1971; Motolinía 1903; Pasztory 1974; Sahagún 1951). Many calendric rituals associated with these Gods occurred prior to planting, either before or at the beginning of the rainy season *rather than following harvest*. The Aztecs, for example, dedicated a feast to Tlaloc during the month Tozoztontli, which corresponds to April, during which flowers, food, and captives were offered. The Cantares Mexicanas describe Tlalocan, the abode of the Tlalocs as a flowery garden paradise (Bierhorst 1985). Rituals to the Tlalocs continued until at least the month Etzalcualiztli, corresponding to June (Broda 1971, 1991: 96; Sahagún 1951). Many of these rites occurred at water shrines in the Basin of Mexico, perhaps similar to

Fig. 7.6 Tlaloc figurine from shrine

the shrine discussed in this paper. The Tlaloc figurines may be associated with rain God effigy vessels, such as those associated with Tlaloc in divinatory almanacs like the Codex Borgia (see Boone 2007: 145).

The previous month, Tlacaxipehauliztli, saw a series of feasts in honor of Xipe Totec. During this month many captives taken in war were killed. Their skins were flayed and their bodies dismembered (Sahagún 1951: 46). Such sacrificial acts remind us of the shrine's burials. This feast was also called "The Offering of Flowers" given the number of flowers offered during this time. Sahagún (1951: 98) recorded how sacrificial victims were often adorned with flower wreaths on their heads and with garlands around their necks, a practice that could explain the flower pollen in the burials. His Nahua informants observed: "I offer him flowers…I provide one with a flower necklace. I place a garland on one…I cover him with flowers…I destroy him with flowers" (Sahagún 1963: 215).

Concluding Thoughts

It really is not difficult to open the pages of historic or ethnographic texts in Mesoamerica and find rich meanings associated with ritual paraphernalia, especially plants and animals. It is rather easy to selectively pull upon those tidbits of information that allow us to not only expand on the general symbolism of ritual items but perhaps also to specify the time of year in which rituals occurred and their associated deities. However, this method is problematic. True, many rituals had very specific and often strict protocols (Boone 2007), a fact that predicts highly structured ritual deposits in general (i.e., Richards and Thomas 1984). But an incredible degree of overlap existed between rituals and in the required offerings to deities, which impedes our pursuit of that eureka moment when we can map the historical account directly onto the archaeological record. Furthermore, this approach would tread rather clumsily through history, grasping at superficial and highly decontextualized similarities as if they unproblematically represented real points of cultural continuity, in this case between periods of time centuries apart.

Maize, beans, amaranth, incense and flowers, especially *cempoalxochitl*, were offered throughout yearly cycles for deities in both solar and ceremonial calendars. In much of Mesoamerica deities possessed few unique attributes. Andrews and Hassig observed: "their qualities, provenances, and powers tended to be diffuse, shared by many Gods" (Andrews and Hassig 1984: 10). Each deity often had multiple names or identities that could be expressed via other means, such as days or years, temporal moments often also shared by other supernatural entities (Boone 2007; Caso 1959). This complexity resists the pursuit not only of singular dimensions of spiritual entities but should also temper efforts to associate specific material objects, such as floral and food offerings, uniquely with one deity, let alone a specific moment of time.

On the other hand, this exercise shows these botanical materials to be more than utilitarian resources constrained only by annual variations in climate. Plants and animals of similar kinds pervade rituals throughout the year. Crops can be stored. Flowers can be dried. Methodologically, then, we should not simply default to contemporary ecological analogues of seasonality to remedy interpretive ambiguity. Pursuing symbolism is intrinsically important. To pull on Lévi-Strauss' old insight, not only animals but plants also are "good to think" both with and through (see Leach 1970: 31–32). They provided the material resources—the raw material—to be set in motion by the conceptual universes of past (and present) people (see Rival 1998). The unique qualities (i.e., liminal) of ritual are established by the active use of resources with unique histories, cycles, associations, and lives (see Vogt 1976). Examining comparative literature sheds light on how plants' multiple and specific dimensions affected the content, temporality, and spatiality of ritual acts. For example, Hill (1992: 127) writes that "the Flower World is the place where the spiritual aspects of living things are found." Flower symbolism pervaded Mesoamerican ritual—a complex that captures the spaces of life, the soul, the heart, death, fire—and ritual paraphernalia in general is often called flowers (e.g., Hill 1992; León-Portilla 1992; Morehart et al. 2005; Vogt 1976).

But where does one end the pursuit of meaning once we decide to move beyond the seemingly obvious ecological characteristics of the "natural" world. My reference to Lévi-Strauss above was, in a sense, deliberate; his analyses of myths and symbolism always retained, as he himself recognized, a degree of initial arbitrariness that can lead to a continual, logical expansion ad infinitum, with one myth's denominator serving as another's numerator (Lévi-Strauss 1975: 136). But this becomes a problem only if our Western need for finality and category demands unambiguous certainty over process and fluidity. For Lévi-Strauss, the ultimate endpoint for the scientist was the human mind. But this route is not essential to enrich our understanding of potentially distinctive universes of thought and practice. The search for meaning itself can be transformative for the scholar by demanding recognition of the textured complexity of different conceptions of space, time, people, places, and things. In other words, tracking the cultural symbolism of ritual helps open our eyes to potential meanings and processes that were not always specific temporally and spatially. Rather, they are thematic, layered, and distributed, pointing us to qualitatively different kinds of spaces and times.

References

Andrews, J. R., & Hassig, R. (1984). Editors introduction: The historical context. In H. Ruiz de Alarcón (Ed.), *Treatise on the heathen superstitions that today live among the Indians native to this New Spain, 1629* (pp. 3–36). Norman: University of Oklahoma Press.

Arnold, P. (1999). *Eating landscape: Aztec and European occupation of Tlalocan*. Niwot: University Press of Colorado.

Benz, B. (1986). *Taxonomy and evolution of Mexican Maize*. Ph.D. Dissertation, Department of Botany, University of Wisconsin–Madison, Madison.

Benz, B. (1994). Reconstructing racial phylogeny of Mexican Maize: Where do we stand. In S. Johannessen & C. A. Hastorf (Eds.), *Corn and culture in the prehistoric New World* (pp. 157–179). Boulder: Westview Press.

Bierhorst, J. (1985). *Cantares Mexicanos: Songs of the Aztecs*. Stanford: Stanford University Press.

Boone, E. (2007). *Cycles of time and meaning in the Mexican Books offate*. Austin: University of Texas Press.

Broda, J. (1971). Las Fiestas Aztecas de los Dioses de la Lluvia. *Española de Antropología Americana, 6,* 245–327.

Broda, J. (1991). Sacred landscape of Aztec calendar festivals: Myth, nature, and society. In D. Carrasco (Ed.), *To change place: Aztec ceremonial landscapes* (pp. 74–120). Niwot: University Press of Colorado.

Carrasco Pizana, P. (1950). *Los Otomíes: cultura e historia prehispánicas de los pueblos mesoamericanos de habla otomiana*. Mexico City: Universidad Nacional Autónoma de México.

Caso, A. (1959). Nombres Calendáricos de los Dioses. *El Mexico Antiguo, 9,* 77–100.

Charlton, T. H. (1970). Contemporary agriculture of the valley. In W. Sanders, A. Kovar, T. H. Charlton, & R. A. Diehl (Eds.), *The Teotihuacan valley project, final report, volume I: The natural environment, contemporary occupation and 16th century population of the valley* (pp. 253–383, Occasional Papers in Anthropology No. 3). University Park: The Pennsylvania State University.

D'Andrade, R. (1995). *The development of cognitive anthropology*. Cambridge: Cambridge University Press.

Diehl, R. A., & Berlo, J. C. (Eds.). (1989). *Mesoamerica after the decline of Teotihuacan, A.D. 700–900*. Washington, D.C.: Dumbarton Oaks.

Douglas, M. (1996). *Natural symbols: Explorations in cosmology*. London: Routledge.

Durán, F. D. (1971). *Book of the gods and rites and the ancient calendar* (F. Horcasitas & D. Heyden, Trans.). Norman: University of Oklahoma Press.

Durkheim, É. (1915). *The elementary forms of religious life*. New York: The Free Press.

Gorenflo, L. J., & Sanders, W. T. (2007). Archaeological settlement pattern data from the Cuautitlan, Temascalapa, and Teotihuacan Regions, Mexico. Occasional papers in anthropology, Number 30, Department of anthropology, The Pennsylvania State University.

Harris, M. (1966). The cultural ecology of India's sacred cattle. *Current Anthropology, 7,* 261–276.

Harris, M. (1979). *Cultural materialism: The struggle for a science of culture*. New York: Vintage Books.

Hastorf, C., & Johannessen, S. (1991). Understanding changing people/plant relationships in the prehispanic Andes. In Robert W. Preucel (Ed.), *Processual and postprocessual archaeologies: Multiple ways of knowing the past. Center for Archaeological Investigations, Occasional Paper 10*. Carbondale: Southern Illinois University.

Hill, J. H. (1992). The flower world of the Old Uto-Aztecan. *Journal of Anthropological Research, 48,* 117–144.

Keber, E. Q. (1995). *Codex Telleriano-Remensis: Ritual, divination, and history in a pictorial Aztec manuscript*. Austin: University of Texas Press.

Leach, E. (1970). *Claude Lévi-Strauss*. New York: Viking Press.

León-Portilla, M. (1992). *Fifteen poets of the Aztec world*. Norman: University of Oklahoma Press.

Lévi-Strauss, C. (1975). *The Raw and the Cooked: Introduction to a Science of Mythology* (J. Weightman & D. Weightman, Trans.). New York: Harper Colophon Books.

Lucas, G. (2005). *The archaeology of Time*. New York: Routledge.

McNeil, C. (2006). *Maya Interactions with the natural world: Landscape transformation and ritual plant use at Copan, Honduras*. Ph.D. Dissertation, Department of Anthropology, City University of New York,

Minnis, P. (1981). Seeds in archaeological sites: Sources and some interpretive problems. *American Antiquity, 46,* 143–152.

Montúfar Lopez, A. (2006). La flora sagrada de Tenochtitlan: Propuesta para una nueva sala del Museo del Templo Mayor. In L. López Luján, D. Carrasco, & L. Cué (Eds.), *Arqueología e Historia del Centro de México: Homenaje a Eduardo Matos* (pp. 493–506). Moctezuma: Instituto Nacional de Antropología e Historia.

Morehart, C., & Morell-Hart, S. (2015). Beyond the ecofact: Toward a social paleoethnobotany in Mesoamerica. *Journal of Archaeological Method and Theory, 22,* 483–511.

Morehart, C. T. (2010). *The archaeology offarmscapes: Production, place, and the materiality of landscape at Xaltocan, Mexico*. Ph.D Dissertation, Department of anthropology, Northwestern University, Evanston, IL.

Morehart, C. T., Lentz, D. L., & Prufer, K. M. (2005). Wood of the Gods: The Ritual use of pine (*Pinus* spp.) by the ancient lowland Maya. *Latin American Antiquity, 16*(3), 255–274.

Morehart, C. T., Meza Peñaloza, A., Serrano Sánchez, C., McClung de Tapia, E., & Ibarra Morales, E. (2012). Human sacrifice in the Northern Basin of Mexico during the epiclassic period. *Latin American Antiquity, 23*(4), 426–448.

Motolinía, T. (1903). *Memoriales*. Madrid: Librería de Gabriel Sánchez.

Parsons, J. R., Gorenflo, L. J., Parsons, M. H., & Wilson, D. J. (2008). *Prehispanic settlement patterns in the Northwestern Valley of Mexico: The Zumpango Region*. Ann Arbor: Museum of Anthropology, University of Michigan Memoirs.

Pasztory, E. (1974). *The iconography of Teotihuacan Tlaloc. Studies in precolumbian art and archaeology* (No. 15). Washington, D.C.: Dumbarton Oaks.

Rappaport, R. (1968). *Pigs for the ancestors: Ritual in the ecology of a new Guinea people*. New Haven, CT: Yale University Press.

Rappaport, R. (1979). *Ecology, meaning, and religion*. Richmond, VA: North Atlantic Books.

Richards, C. C., & Thomas, J. S. (1984). Ritual activity and structured deposition in later Neolithic Wessex. In R. Bradley & J. Gardener (Eds.), *Neolithic studies* (pp. 177–188). Oxford: British Archaeological Reports.

Rival, L. (Ed.). (1998). *The social life of trees: Anthropological Perspectives on tree symbolism*. Oxford: Berg.

Robb, J. E. (1998). The archaeology of symbols. *Annual Review of Anthropology, 27,* 329–346.

Roscoe, P. B. (1995). The perils of 'positivism' in cultural anthropology. *American Anthropologist, 97,* 492–504.

Ruiz de Alarcón, H. (1984). *Treatise on the heathen superstitions that today live among the Indians native to this New Spain, 1629*. Norman: University of Oklahoma Press.

Sahagún, B. (1951). *Florentine codex, general history of the things of New Spain: Book 2, the ceremonies*. Santa Fe, NM: School of American Research.

Sahagún, B. (1963). *Florentine codex, general history of the things of New Spain: Book 11, earthly things*. Santa Fe, NM: School of American Research.

Sanders, W. T., Parsons, J. R., & Santley, R. S. (1979). *Basin of Mexico: Ecological processes in the evolution of a civilization*. New York: Academic Press.

Santamaría, M. (1912). *Las Chinampas del Districto Federal*. Mexico: La Secretaria de Fomento.

Turner, V. (1969). *The ritual process: Structure and anti-structure*. New Brunswick, ME: Aldine Transaction.

Vogt, E. Z. (1976). *Tortillas for the Gods: A symbolic analysis of zinacanteco rituals*. Cambridge: Harvard University Press.

Wellhausen, E. J., Roberts, L. M., & Hernandez, E. X. (1952). *Races of Maize in Mexico: Their Origin, characteristics and distribution*. Cambridge: Bussey Institution of Harvard University.

Chapter 8
A Compelling Intersectionality: Paleoethnobotany, Social Theory, and Feminist Commitments

Margaret W. Conkey

Introduction

The term "intersectionality" was coined (Crenshaw 1991) and subsequently developed and used to characterize approaches that explored the interconnections between systems of oppression (e.g., Hull et al. 1982). But it has also come to refer to several other types of relationships, none of which are mutually exclusive (Collins 1999). These varied intersectional relationships can readily be applied to contextualizing archaeological practices: (1) that ideas and the social structure within which they occur are interrelated; (2) that we can ask, within a discipline, with its own ideas and social structures, how our own subsets and perspectives intersect? And (3) that the hierarchies of gender, race, economic class, sexuality, nationality, ethnicity, and so forth, are intersected; there is an intersectedness of these hierarchies. This last set of relationships is the more usual domain to which the term "intersectionality" refers, especially in the literatures of feminist scholarship.

We in archaeology have experimented with at least the second of the above types of intersectionality (e.g., Conkey 2005, between feminist and indigenous archaeologies; Franklin 2001; Battle-Baptiste 2011; Sterling 2015 between black feminist thought and archaeologies). Others have often developed understandings of how our ideas and contextual social structures are intertwined if not co-constituted (e.g., Leone 2010; and many others). While the above is admittedly a rather rough and simplistic characterization of what "intersectionality" is about or how it has impacted the understandings of social phenomena, the concept that there are analytical and interpretive "positions" that intersect in dialectical, productive, and mutually constitutive ways has the potential to frame archaeological research and to provide, if you will, a kind of archaeology of archaeology.

M.W. Conkey (✉)
Department of Anthropology, University of California, Berkeley, USA
e-mail: meg@berkeley.edu

M.P. Sayre and M.C. Bruno (eds.), *Social Perspectives on Ancient Lives from Paleoethnobotanical Data*, DOI 10.1007/978-3-319-52849-6_8

In this chapter, I will take the concept of intersectionality out of its more usual domain (relationships of the third type) and use it to provide a somewhat different understanding of how some archaeological research today can work (more in line with relationships of the second type). In particular, I want to use the examples, on the one hand, of the field of paleoethnobotany—a key theme of this volume—and, on the other hand, of the elaboration of social theories and especially feminist theory and practice to illuminate how a specific intersectionality among them has been made manifest and can contribute to more robust and richer understandings of aspects of the human past. While the research trajectory of Christine Hastorf provides one of the best individual examples of this particular intersectionality, there are other recent studies and approaches that are nudging the field of paleoethnobotany in this direction.

Like most accomplished paleoethnobotanists, Hastorf is a very strong defender of the power and role of plants in past human life and of the centrality of botanical analyses in any archaeological inquiry. But her own paleoethnobotanical work as a methodological and interpretive expertise has long been situated in an embedded structure of a dynamic and often original use of social theory, especially when working with plants. Much of her work, especially since the 1990s, has perhaps set the stage for the intersectionality being explored here in that she has consistently drawn upon various themes and theoretical approaches in social theory, many of which can be linked to some basic epistemic principles derived from feminist theory and feminist practice. The importance of framing this work as an intersection of at least two variable and shifting dimensions of research is to remind us that we do not just "apply" theory nor just "do" methods, and that there is a creative and interpenetrating relationship that can yield genuinely original insights.

Of course, many would unhesitatingly begin an inquiry into a more engendered approach in paleoethnobotany with the 1991 Hastorf contribution to the edited volume, *Engendering Archaeology—Gender, Space and Food in Prehistory*. In that same volume (Gero and Conkey 1991) is another significant and mobilizing chapter—by Watson and Kennedy (1991)—that is also a hallmark paper for not just "considering" gender in a paleoethnobotanical study but for the incorporation of feminist principles and critical analysis of heretofore unquestioned assumptions (in this case, about the role of women in the much-heralded "invention" of agriculture). A close inspection of some subsequent research publications and research practices, albeit by still a relatively few scholars, reveals what can be considered feminist practice and congruent with feminist principles and theory, approaches that are themselves, "beyond gender" (Engelstad 2007).

Where I want to go with this is to take three tacks: first, I will indeed say a few things about some of the work by Hastorf and how she has fundamentally generated a paradigm for what one might call a "feminist-inspired social paleoethnobotany", some of which is only emergent in the past few years. One could track her work in some detail over the past 20 + years in particular and a feminist historian of science would be able to show the specific connections, links, and intellectual moves that have generated her *oeuvre* (or corpus of work).[1] Rather, for this short consideration, her publication history shows that she has engaged with all sorts of key concepts of social theory and cultural *praxis* openly, even in the very titles for her publications.

Second, I will refer to a set of what have been called feminist principles for how to "do" one's work as a feminist: what epistemological and action principles inform one's work "as" a feminist. Third, I will refer to several recent publications and studies that have added significantly to the corpus of paleoethnobotanical work that not only contribute to a more social paleoethnobotany [as we have seen in a more social zooarchaeology (Russell 2012) or a more social bioarchaeology (Agarwal and Glencross 2011)] but point toward a more genuinely intersectional paleoethnobotany that takes up one or more of the feminist principles that have to be described.

Considering Hastorf's Scholarship

First, "theory" is not merely some framework or grid to map onto the data or particular archaeological situation, nor is it a singular and thus repetitive approach. Rather, it seems that theory, and in the specific case here of Hastorf's work, is something that is revelatory (Culler 1994); something that allows her (and therefore, allows us) to see things in a new way, and that expands our views and our understandings rather than just confirms the extant. She consistently sees how certain theoretical assumptions—even if developed for a very different set of cultural materials or contexts—can illuminate the world of the Andean past that she has come to understand so well. I note here, for example, one of her recent approaches to households and foodways in the Andes was mobilized by the work of the anthropologist Gell (1998) who developed and drew upon several concepts—such as "distributed persons" and "art as agency"—to reframe the anthropological study of art: concepts that Hastorf creatively and successfully adapts to and enlists in her understandings of households and their dynamics of intimacy and social relations (Hastorf 2009).

While there is no doubt that Hastorf is also very much the lab scientist, who is concerned with such issues as sampling, methodologies, experiments, equipment, and has even written about laboratory goals (e.g., Coil et al. 2003; Hastorf and Wright 1999), it is even in her laboratory practices that feminist principles are at work. That is, her uses of theory are not "just" theoretical, not just to weave a better or more nuanced tale of humans, culture, and plants.

I have gone through the titles of Hastorf articles and book chapters—a major task since there are close to 100 of them!—from the perspective of a word sleuth or

[1]I have recently tried to do this for another archaeology scholar, Olga Soffer, showing how, on the one hand, she moved into a more social and even feminist engagement with her Paleolithic archaeology research and yet how, on the other hand, we could see its roots in her earlier publications and interests (Conkey 2008).

what you can find now on the Internet as Textalysers or Text Content Analysis. That is, what are key words in her research that are, furthermore so important that they are in the titles (and, to be sure, as a major feature of the text itself). The anthropological concepts and terms that pervade her titles include: cultural life; history; cultural implications; community; community-making; social memory; distributed persons; cultural interpretation; *habitus*; contextual; style; tradition; identity; ceremonial commensality; social and political aspects of feasting, of plant use and agriculture; food and society; foodways; households; food for the ancestors, deities, and elites; gender, space and food; plant/people relationships; homes and stores (storage); empires; heads of state; power; politics; political inequality exchange; ritual; ritual performance, liturgy and meaning; labor. Most of the publications with such titles definitely include fine-grained botanical analyses (as well as of other material evidence) with reference to methods and empirical documentation, yet these are never standing alone, but embedded in—and made relevant because of—a certain anthropological/theoretical problem and a well-defined justification for why such a problem orientation not only might make sense but how it amplifies the interpretive possibilities for the data at hand.

While there should be little doubt that if one gave this list of terms and concepts to someone to affiliate the articles with a discipline that the affiliation would be to anthropology, it would be almost impossible to guess immediately (with few exceptions) that these are almost all articles about plants, agriculture, cooking, economy, or ecology. Additional terms that feature might allude to a more delimited domain of social inquiry: women's labor, domestic economy, the cook's perspective, midwifery-as-metaphor, women and children first, households. But it is not "just" these latter that qualify for a feminist label. Rather, it is what Hastorf does with all of these concepts and approaches, even if there is no mention of "women" or "gender" (a term that does not, in fact, feature often in her work) that provides the link with the feminist practice.

How to Do Our Work "as a Feminist"

So maybe we should turn to what is meant by feminist practice or by "doing archaeology as a feminist." I draw here from the important programme put forth by epistemologist/philosopher of science, Longino (1994), which was commented upon and elaborated by Wylie (1995): "In Search of Feminist Epistemology" or as Wylie has put it: "what does it mean to do (social) science as a feminist?" In an elegant attempt to reframe the question that has befuddled many a feminist since the 1970s, now known as "the feminism question in science," or, "is there a feminist method?", Longino turned the issue on its head and advocated that the question be more about how one practices their field of inquiry AS a feminist, more so than that there is a uniquely feminist method. This issue was taken up by some of us for archaeology in a special issue of the *Journal of Archaeological Method and Theory* (Conkey and Wylie 2007), although here I want to go back to Longino's initial principles against which to evaluate or reflect upon not just Hastorf's work, but the implications for the more specified question: "what would/does it mean to do paleoethnobotany as a feminist?".

But first a warning or at least a qualification: where we are going is about—to paraphrase Ericka Engelstad —"much more than gender" (Engelstad 2007). Indeed, while feminist inquiry—even in archaeology—has become much more than gender, what these Longino-inspired feminist principles "do" is (via what she calls "theoretical virtues") to "prevent gender from being disappeared"; they "make gender a relevant axis of investigation," and above all, they should not just allow for but positively enhance the likelihood that our investigations will "complicate" our all-too-often essentialized grand narratives. As Hastorf herself has written: foodways, to her, are an entry into lived social relations, into home life, and at a scale that is more "intimate" than the "often broad sweeps archaeologists have traditionally used" (2009: 10). Yet, as discussed in one of her most thorough overviews (Hastorf 2001) of what a valuable multi-scalar approach can be in archaeology and especially through plant remains (visible or not!), she demonstrates how that "intimate" inquiry can and should be balanced with asking the larger scale questions, those of the "broad sweeps." The question to ask more widely of paleoethnobotany is if research has gone "beyond gender?"

There are two directions in which to go: on the one hand, there are Longino's principles or what she calls "theoretical virtues" that I am going to list but not elaborate upon. More substantively, I will take the implications of these, especially as developed by Wylie (1995; see also Wylie 2007a) for archaeology.

Longino's theoretical virtues are: (1) empirical adequacy; (2) novelty (such as a departure from entrenched assumptions): (3) ontological heterogeneity (a phrase that only an epistemologist could love—or even throw out there!) in two senses: to "prefer theories that take account of individual difference among the objects of study" and to" reject [those] theories of inferiority in which differences are ranked and/or diverge from a norm" [and thus are] treated as a deviance or failure, and thus allow "equal standing for different types" and investigate the difference; (4) complexity of relationship; a preference for theories that treat "complex interaction as a fundamental principle of explanation." More specifically, "replace simple models of single factor control in social contexts with more complex models of social interaction"; (5) the applicability to current human needs; e.g., to acknowledge and pursue the potential role of scientific knowledge in improving the material conditions of human life; (6) a diffusion of power, which is an expansion of No. 4 (above), that is, pursue practices that are "widely distributed".

For those who know Hastorf's work, one can see how it can readily be used as examples of these principles: it is not "just" the most locally available firewood in the Andes that is preferred; it is more complicated than that, due to mythological and symbolic understandings, including the ideological world view of the conquering Inca and the subsequent differential uses of wood between households of different statuses (e.g., Hastorf et al. 2006); it is not "just" the male-focused *chicha* beer drinking ceremonies that solidified or reinforced the Inca state, but the labor of women in corn processing for the *chicha* that made it all possible (Hastorf 1991);

households are not just places where daily life plays out or where economics are at work, but they are a "stand in for the social relations required to construct the home and then to live within it"..."[B]oth structure and agency [are] harnessed in this study of households" (Hastorf 2009).

From the Longino theoretical virtues, one can develop, thanks to Wylie (1995, 2007a, b), a set of guidelines, a sort of a checklist for feminist practice. This would not, I hasten to add, be necessarily limited to self-identifying feminists; rather, as feminists ourselves would say, these are practices that make for a better, stronger, more honest, and more productive archaeology in general! The challenge, as Wylie has so clearly articulated, is how to design an archaeological practice that is cognizant of its own partiality; recognizes yet pushes at its limitations; makes full use of its potential as an empirical discipline to not only yield new insights/ideas/things about the human cultural past, but also, to do so in an number of ways that necessarily disrupt, or at least problematize, apparently "settled assumptions"/established taken-for-granteds. One can see how various works in paleoethnobotany from the "start" (e.g., Hastorf 1991; Watson and Kennedy 1991) would "satisfy" such suggestions.

From these, one might develop a subsequent set of questions to ask ourselves in feminist-inspired research: Have we always acknowledged human agency? Have we resisted pressures to simplify, to reduce the complexity of the cultural subjects of concern and interest? Do we openly mistrust practices that reinforce or sustain simplistic accounts and certainty, while recognizing and engaging with ambiguity (Gero 2007) and contingency? Have we brought not just multiple but divergent perspectives to bear? And above all, have we respected the very "materiality" of the archaeological record—those "evidential constraints" (Wylie 1992)? Have we put a high priority on ways of doing research, on dealing with communities and institutions that foster more equality of intellectual authority, democratize the process of inquiry?

Recent Studies in a "Social" Paleoethnobotany

Several notable studies in paleoethnobotany have appeared, especially in the past 10–12 years that are, in varied ways, strong candidates for work that can be suggestively carried out as if from a perspective of such feminist "theoretical virtues" and feminist inspired practice. Of particular note is the 1999 article by Gayle Fritz, who, in some ways, took up the challenge of Watson and Kennedy (1991) on the issue of the origins of agriculture in the North America and the role of women and men in that much—heralded transformative process. But Fritz's well-known study of the domestication of gourds is not simply an attempt to show that either men or women were the responsible agents. Rather, this is a superb example of an analysis that complicates the story by suggesting that it is not a question of "either/or" or "who" but a story of situational use of some plants, probably not as intentional domesticates for subsistence, but of an association of the gourds with a fishing economy, their utility in the use of nets and, as well, not linked at all to the later domestication of the classic crops of corn and beans. In an admirably feminist challenge, Fritz—supported by detailed paleoethnobotanical and archaeological analyses—shows how the limited bipolar notions of men or women as the relevant actors in a cultural process are to be questioned and are limiting in advancing our understanding of a culturally and environmentally embedded process that, while not "disappearing gender" nonetheless shows that the history was more contingent, and more situationally generated than a simplistic model that considers all domesticates as a package and all actors as defined by an inflexible division of labor. In some ways, despite its relatively early appearance in the more engendered archaeological literature beginning in the 1990s, Fritz's study is perhaps the most robust and feminist-informed to date.

But into the twenty-first century, numerous studies appear that attest that paleoethnobotanists are "thinking gender" and "thinking like a feminist": the intersectionality is, if you will, taking hold. Among the publications to note include the approach by Morehart and Helmke (2008) that examines Late Classic Maya gender and social relations but explicitly rejects simplistic gender roles and decontextualized inquiries into what is a complex political ecology of productive labor. They compare two very different sites and show how making generalizations based on reductionist notions of labor are not viable, while implicating gender in both ritual and political ecology in nuanced and varied, situational ways. A second contribution of the early 2000s is one of my favorites, the study of elderberry use among the Tsimshian of the Northwest coast in both pre- and post-contact contexts that combines intra-site spatial analyses, and detailed botanical analyses but in the specific context of dynamic and changing political situations, finding that women and elderberry production was actually on the increase after European contact, at least for a while (Martindale and Jurakic 2004). Their work is markedly feminist in that it queries issues of power, shows heterogeneity in the story of contact situations, asks about the materiality of women's and men's work, produces an account that brings together multiple lines of evidence, including the botanical, and yet raises on-going questions in what is clearly a reframing of the culture contact situation and how it played out in the daily lives of the people impacted.

Another analysis of relevance is that of VanDerwarker and Detwiler (2002) that challenges the long-standing simplistic assumption that associates women with the private/domestic and men with the public/political. In the case of the Coweeta Creek site (eighteenth century) in the US Southeast, they show that food processing, much of which is known to be that by women, was carried out adjacent to the public townhouse and thus, very visible, very public and integral to the on-going-ness of public and political life. The activities they document are also in a contact historical context that prompts the authors to engage with a more decolonized approach and to call for this to be further developed in such archaeological settings.

VanDerwarker later joined (VanDerwarker et al. 2016, online in 2015) with many colleagues to publish a substantive review of New World Paleoethnobotany as it has been developing in the twenty-first century so far. This article covers many key topics in the field of paleoethnobotany as practiced in the "new World," and an end section specifically addresses the topic of gender in concert with the topics of "Identity and Culture Contact." While there is little here in terms of specific analyses or analytical frameworks, nor a call for explicitly feminist approaches, it is relevant to the paleoethnobotany community to be reminded of such topics as part of what a research agenda could and should be. As this piece was being written, Morehart and Morell-Hart (2015, online in 2013) took up the need to argue for a more social paleoethnobotany with a synthesis, from their perspectives, on current research that could fall under this rubric.

This is an honest assessment whereby they recognize that much needs to be done to move the field toward what I am calling here a genuine intersectionality, even one of "the social" and the archaeobotanical. They note that human–plant relationships can shed light on power, agency, social structures, and normative constraints but this has been carried out, so far, primarily with respect to issues of ritual and political ecology (as in Morehart and Helmke 2008). That is, there needs to be a push toward additional pathways, they suggest, toward a social paleoethnobotany through the considerations of materiality, past gender relations and environmental knowledge as well as a more sophisticated consideration of "the effect of scale" on analysis and interpretations. They are honest in their assessment that most of the work to date has been along the lines of what Hegmon (2003) has called "processual-plus" that tends, it seems, to not yet embrace some of the more reflexive demands of the feminist programme as described above. Given their own research foci in Mesoamerica, they are especially concerned about the "dearth" of social paleoethnobotanical research in that area, especially given the very rich literature in archaeology on gender and with feminist excellence in ancient Mesoamerica.

Foundational to the Intersectionality of a Social/Feminist Paleoethnobotany

Thus, there are glimpses here and there into what a more fully developed intersectional social/feminist paleoethnobotany can be from a recent generation of researchers who are looking over their field and its potentials. Some of the above cited studies are definitively paving the way. Yet I would have to argue, after having reviewed some of the field, that it is still in the work of Hastorf that one can see the most elaborated and expanded ways in which this intersectionality can be/has been played out. As with most feminist-inspired research, inquiry is never final or totally finished: there are not only new methods to, as Hastorf (2001) has written, "make the invisible visible," to bring out those "hidden jewels" of archaeology. But there are more questions to ask, past answers to go back to and rethink, and new approaches to bring to bear. As part of the feminist ways of doing, indeed, such provisionalism and constant revision are core.

In fact, in returning to Hastorf's work, we can identify some precise ways in which it meets and exceeds some of these expectations and hopes for research practice and interpretation. It is still an aspirational model for the field and the researchers who want to chart a more ampliative and intersectional pathway. In her work, theory is not just something to which to appeal at the beginning of a paper, chapter, review, or book. It is a guiding thread that is in dialogue with the empirical evidence at hand; it is itself a shifting set of principles and practices that are more revelatory than containing. Yet discussions of theory are straightforward, neither opaque nor mystifying. In Hastorf's work, she does not dwell on theory in abstract terms but sets about to see how it can work in the context of the particular issues she is addressing. She is not one to quickly borrow the latest theoretical fad, but she mines on-going theoretical developments deeply and often finds ways in which such approaches can provide yet another vantage point on issues that are key to her overarching goals of making more, making deeper and often more complex and nuanced, sense of people and plants, of the on-the-ground human activities and meaning-making that mobilize culture and precipitate the events to which archaeologists have some access.

For feminists and others, this is what theory should be about: to open up, to expand, to reveal something new, rather than to contain, constrain, and perpetuate the same narrative (as, for example, with some strict Freudians, Marxists, or Evolutionary Psychologists), despite the wider misunderstanding that what matters to feminists is only "being about women." That our explanations should be multicausal and not mono-causal, and that pluralism should prevail in scientific inquiry (Chang 2012) makes sense to a practicing feminist, who understands the complexities, the intersectionalities, and that even core concepts are complicated and demand nuanced attention. As a feminist, Hastorf is perpetually curious; a hallmark of feminist thought as elegantly elaborated by Cynthia Enloe (*The Curious Feminist* 2004) and yet, to use one of Gell's (1998) own terms, Christine is "enchanted" by plants and plants in human life. And it is we who are the richer for it.

Conclusion

To conclude, it is relevant to ask "what is to be gained from such intersectionalities —between paleoethnobotany, social theory and the feminist practice of archaeology?" This should need little development here for if one steps back and considers that from this corpus of "Hastorfian" work and from the new directions toward a more social and even more "feminist-inspired" paleoethnobotany, we have genuinely expanded our understanding and knowledge of a fuller range of both past human lives and of how one can "do" archaeology. This is especially striking given that our wider audiences are often so concerned with what they perceive to be archaeologically available: limited evidence, lack of informants, and/or lack of texts. How can the flotation of archaeological soils actually "reveal" anything about social relations or community-making? How can microscopic analyses of starch grains allow inferences about differential status and access to resources? How can the spatial analysis of plant processing make manifest differential gender roles in past societies? These are but some of the remarkable questions that have been or can be addressed, when framed with theoretical approaches and concepts that demand an engagement with the everyday and the intimate as well as with the "broad sweeps," an engagement with the inequalities in past lives, with strategies for daily practice, and with the complexities and complications that surely characterized past human lives and societies, as they still do for us today. The intersectionality approach renders these complications to be real, and reveals that power dynamics have always been at work. Thus, to take a critical stance on how archaeological knowledge is produced and what kind of knowledge it is or tends to be is both a hallmark of the feminist practice and characteristic of what this feminist-inspired paleoethnobotany can be. We are often jarred into seeing things differently and through another lens, which should only expand understanding—even if it is more complicated—and not reduce it by containment or by delimiting it. Some might say that the presence of a seed (and its identification) in an archaeological context alone will "speak" to us, but when in dialogue with a revelatory theoretical framework, that seed will actually shout.

References

Agarwal, S., & Glencross, B. (2011). *Social bioarchaeology*. New York: Wiley.

Battle-Baptiste, W. (2011). *Black feminist archaeology*. Walnut Creek, CA: Left Coast Press.

Chang, H. (2012). *Is water H_2O? Evidence, realism and pluralism*. New York: Springer.

Coil, J., Korstanje, M. A., Archer, S., & Hastorf, C. A. (2003). Laboratory goals and considerations for multiple microfossil extraction in archaeology. *Journal of Archaeological Science, 30,* 991–1008.

Collins, P. H. (1999). Moving beyond gender: Intersectionality and scientific knowledge. In M. M. Ferree, J. Lorber, & B. B. Hess (Eds.), *Re-visioning gender* (pp. 261–284). Thousand Oaks, CA: SAGE.

Conkey, M. (2005). Dwelling at the margins, action at the intersection? Feminist and indigenous archaeologies, 2005. *Archaeologies, 1*(1), 9–59.

Conkey, M. (2008). One thing leads to another: Gendering research in archaeology. In L. Schiebinger (Ed.), *Gendered innovations in science and engineering* (pp. 43–64). Stanford, CA: Stanford University Press.

Conkey, M., & Wylie, A. (Eds.). (2007). *Journal of Archaeological Method and Theory. Special Issue: Doing Archaeology as a Feminist 14*(3), 209–358.

Crenshaw, K. (1991). Mapping the margins: intersectionality, identity politics and violence against women of color. *Stanford Law Review, 43,* 1241–1299.

Culler, J. (1994). Introduction: What's the point? In M. Bal & I. Boer (Eds.), *The point of theory: practices of cultural analysis* (pp. 13–17). Amsterdam: Amsterdam University Press.

Engelstad, E. (2007). Much more than gender. *Journal of Archaeological Method and Theory, Special Issue: Doing Archaeology as a Feminist, 14*(3), 217–234.

Enloe, C. (2004). *The curious feminist: Searching for women in the new age of empire*. Berkeley and Los Angeles: University of California Press.

Franklin, M. (2001). A black feminist-inspired archaeology? *Journal of Social Archaeology, 1*(1), 108–125.

Fritz, G. J. (1999). Gender and the early cultivation of gourds in Eastern North America. *American Antiquity, 64*(3), 417–429.

Gell, A. (1998). *Art and agency: An anthropological theory*. Oxford: Clarendon and Oxford University Press.

Gero, J. (2007). Honoring ambiguity/problematizing certitude. *Journal of Archaeological Method and Theory, Special Issue: Doing Archaeology as a Feminist, 14*(3), 311–327.

Gero, J., & Conkey, M. (Eds.). (1991). *Engendering archaeology: Women and prehistory*. Oxford: Blackwell Publishers.

Hastorf, C. A. (1991). Gender, space and food in prehistory. In J. Gero & M. Conkey (Eds.), *Engendering archaeology: Women and prehistory* (pp. 132–159). Oxford: Blackwell Publishers.

Hastorf, C. A. (2001). Making the invisible visible: The hidden jewels of archaeology. In P. Drooker (Ed.), *Fleeting identities: perishable material culture in archaeological research* (pp. 27–42, Center for Archaeological Investigations Occasional Paper 28). Carbondale, IL: Southern Illinois University Press.

Hastorf, C. A. (2009). *Houses, feasts, and distributed people in the later prehistory of the Sausa of the Central Andes*. Paper presented at the Household Archaeology Conference, University of Utah, Salt Lake City, UT, February 2009.

Hastorf, C. A., Whitehead, W. T., & Johannessen, S. (2006). Late prehistoric wood use in an Andean Intermontane Valley. *Economic Biology, 59*, 337–355.

Hastorf, C. A., & Wright, M. (1999). Interpreting seeds contained in dung from archaeological flotation samples: A charring experiment from the Andes. *Journal of Ethnobiology, 18*(2), 211–227.

Hegmon, M. (2003). Setting theoretical egos aside: Issues and theory in North American Archaeology. *American Antiquity, 68*(2), 213–243.

Hull, G. T., Scott, P. B., & Smith, P. (1982). *All the women are white, all the blacks are men, but some of us are brave*. Westbury, NY: Feminist Press.

Leone, M. (2010). *Critical historical archaeology*. Walnut Creek, CA: Left Coast Press.

Longino, H. (1994). In search of feminist epistemology. *The Monist, 77*(4), 472–485.

Martindale, A., & Jurakic, I. (2004). Northern Tsimshian elderberry use in the late pre-contact to post-contact era. *Canadian Journal of Archaeology, 28*, 254–280.

Morehart, C., & Helmke, C. G. B. (2008). Situating Power and Locating Knowledge: A Paleoethnobotanical Perspective on Late Classic Maya Gender and Social Relations. In E. Brumfiel, & C. Robin (Eds.), *Gender, households and societies: Unraveling the threads of the past and the present. Archaeological Papers of the American Anthropological Association* (Vol. 18, Issue No: 1, pp. 60–75). New York: Wiley-Blackwell.

Morehart, C., & Morell-Hart, S. (2015). Beyond the ecofact: toward a social paleoethnobotany in Mesoamerica. *Journal of Archaeological Method and Theory, 22*, 483–511.

Russell, N. (2012). *Social zooarchaeology: humans and animals in prehistory*. Cambridge: Cambridge University Press.

Sterling, K. (2015). Black feminist theory in prehistory. *Archaeologies, 11*(1), 93–120.

VanDerwarker, A., Bardolph, D., Hoppa, K., Thakar, H. B., Martin, L. S., Jaqua, A. L., et al. (2016). New World paleoethnobotany in the new millenium (2000–2013). *Journal of Archaeological Research, 24*(2), 125–177.

VanDerwarker, A., & Detwiler, K. (2002). Gendered practice in Cherokee foodways: A spatial analysis of plant remains from the Coweeta Creek Site. *Southeastern Archaeology, 21*(1), 21–28.

Watson, P., & Kennedy, M. (1991). The development of horticulture in the Eastern Woodlands: Women's role. In J. Gero & M. Conkey (Eds.), *Engendering archaeology: Women and prehistory* (pp. 255–275). Oxford: Blackwell.

Wylie, A. (1992). The interplay of evidential constraints and political interests: Recent archaeological research on gender. *American Antiquity, 57*(1), 15–35.

Wylie, A. (1995). Doing philosophy as a feminist: Longino on the search for a feminist epistemology. *Philosophical Topics, 23*(2), 345–358.

Wylie, A. (2007a). Doing archaeology as a feminist: Introduction. In M. Conkey & A. Wylie (Eds.). *Journal of Archaeological Method and Theory, Doing Archaeology as a Feminist, 14,* 209–216.

Wylie, A. (2007b). The feminism question in science: What does it mean to 'do social science as a feminist'? In S. Hesse-Biber (Ed.), *Handbook of feminist research* (pp. 567–578). Thousand Oaks, CA: SAGE.

Chapter 9
Commentary: Paleoethnobotany Beyond Diet, Environment, and Ecology

Deborah M. Pearsall

> While contemporary palaeoethnobotanists often engage with so-called "processual" issues of diet, environment, and ecology, the research agenda has broadened to include many other themes, such as foodways, gender, interpersonal power issues, identity, trade and exchange, political economy, and all the other topics archaeologists generally engage with through analysis of material culture. After all, for most of human history, the majority of "material culture" was constructed from plants, the most diverse and dynamic group of raw materials available. (Hastorf 2008b: 1790)

> Food is the meeting place of nature and culture. … The many trajectories that food allows us to follow into the past will enrich all of archaeological inquiry and place food archaeology firmly in the center of the discipline over the coming years. (Hastorf 2008a: 1387, 1395)

Introduction

In this chapter I have the pleasurable task of commenting on the chapters in this volume, which highlight the broad research agenda of social paleoethnobotany, or paleoethnobotany beyond diet, environment, and ecology (Morehart and Morell-Hart 2013). My perspective on contemporary practices and approaches in this field is informed by the early and on-going work of Christine A. Hastorf. In each of these chapters, we can follow food into the past in ways that provide new, fresh insights into the interrelationships among plants and people. I write this commentary as a retired paleoethnobotanist who was both an insider and outsider: I was a methods person who studied both macro- and microremains; my research focused largely on diet and environment in the context of early agriculture in the Neotropics. From my perspective three themes resonate strongly in these chapters,

D.M. Pearsall (✉)
Department of Anthropology, University of Missouri, 112 Swallow Hall, Columbia MO 65211, USA
e-mail: pearsalld@missouri.edu

M.P. Sayre and M.C. Bruno (eds.), *Social Perspectives on Ancient Lives from Paleoethnobotanical Data*, DOI 10.1007/978-3-319-52849-6_9

and will be used to structure my commentary: seeking “the invisible”; insights into spatiotemporal characteristics of rituals; and revealing the social realm through paleoethnobotany. I end with some thoughts about the challenges of inferring complex human behaviors from the archaeobotanical record.

Seeking “The Invisible”

In her 2001 book chapter, “Making the invisible visible: The hidden jewels of archaeology,” Christine Hastorf reminds archaeologists that we are missing most of our artifacts (the organic ones) and that the meanings of all artifacts are rarely visible. One way to make the invisible visible is to add different scales of analysis to a project, by analyzing macroremains, pollen, phytoliths, starch, biochemical markers, and so on. These multiple lines of evidence provide different views on the past, and reveal its subtlety and complexity.

Chevalier and Bosquet illustrate this approach in their chapter, which seeks a better understanding of the lives of Middle Belgium’s LBK peoples. Five kinds of archaeobotanical data from Remicourt “En Bia Flo II” and other sites were investigated to understand economies and food choices of pioneer and later LBK populations. Results were interpreted in terms of paleoenvironment (pollen and wood charcoal) and foodways (seeds, phytoliths, starch). I was impressed by the integration of pollen and wood charcoal results: pollen data showed a progressive forest clearing, which wood results confirmed at higher resolution. Importantly, pollen indicated the *extent* of open areas on the landscape, which wood data could not do. The presence of annual weedy species, many identified by macroremains, and the absence of perennial and shrubby species, suggested intensive crop cultivation without long fallowing. From the combined data the authors posited settlement movement after 100–200 years (3–4 generations), providing insights into the social system. Phytolith and starch data revealed plants absent or rare in the macroremains record, such as barley and oat. Lack of charred barley suggested that processing and cooking practices of those foods were likely different from practices for wheat. Studying botanical data that have different pathways into the archaeological record revealed different human behaviors. This chapter demonstrates the potential of different scales of analysis for separating cultural, ecological, and taphonomic influences on patterning in botanical assemblages.

Sites with extraordinary preservation (i.e., waterlogged, dry, ash-covered) provide the outlines of what lifeways and social worlds might have been like at sites characterized by “ordinary” preservation conditions (i.e., through charring) (Hastorf 2001). In her review of dry rock shelter sites in northwestern Argentina, for example, Korstanje was able to compare species variability in edible plants, nonedible plants (fuels), and plant-based artifacts between agricultural and preagricultural occupations. She discovered strong similarities in wild plant use between these occupations, and greater variability in edible wild species used by agriculturalists. The precision of identification possible under these preservation conditions allowed her to assess distances/labor needed to procure resources. On a very different scale of analysis, Farahani and colleagues made use of extraordinary preservation at La Joya de Ceren, El Salvador, to identify culinary stations, and to demonstrate how storage, production, and consumption activities can overlap in a domestic setting. Spatial relationships among culinary objects and food remains revealed "taskscapes," for example, those centered on metates (grinding stones).

How can we make perishable items more prominent in our reconstructions at sites with ordinary preservation? As Hastorf (2001) warns, we should not create just-so stories about the past based on extraordinary sites or ethnography. What she urges is a change of perspective: rather than assuming that what is visible in the excavation is all that there was, assume that the excavated evidence is only the beginning. Imagining the organic things that might have been pushes us to explore ways of seeing them, for example, through biochemical markers, microremains, and associated items that are visible.

Insights into Spatiotemporal Characteristics of Rituals

Ritual practices construct a kind of society that leaves a material record—icons, memorials, architecture, and boundaries between spaces. Studying these provides us with clues to the significance of ritual practices. Today in the Andes of South America, indigenous peoples act within webs of their living and deceased kin, in a landscape that is animated through these interactions (Hastorf 2003, 2007). In her research at the Middle Formative site of Chiripa, Bolivia, for example, Hastorf (2003) drew inferences about shifts in the structure of social organization from changes in burial practices. Association of serving vessels, such as those used in public ritual consumption, with some burials, suggested pits were reopened to feed and tend the central figure of the interment, which changed from female to male over time.

A potentially significant source of complexity in the study of ritual practices is the situation in which the same spaces, objects, and foods are the loci of ritual and daily life. This is the situation explored by Sayre and Whitehead at Conchopata, a Middle Horizon administrative center in Ayacucho, Peru. Spaces at the site were designated ritual or domestic based on architectural criteria, with public feasting defined as the making of maize or *molle* beer, large-scale food preparation, and costly discard of meat cuts. Charred macroremains of five major plant taxa (*Chenopodium, Schinus molle*, parenchyma and tubers, maize, coca) were compared between ritual and domestic contexts in terms of density and ubiquity. No clear separation between ritual and domestic spaces was found, leading the authors to conclude that "(I)n the end, the remains recovered from Conchopata leave us with the impression that domestic and ritual life continuously blended together to leave behind a mixed record of daily activity in an early urban center." The density data illustrate the spatial complexity of activities surrounding ritual and feasting at this site: the highest densities of plants associated with brewing (fuel and maize or *molle* remains) were found in domestic areas, namely spaces adjacent to patio areas of the right size for feasting. Many ritual areas, including the center, had low densities of all charred remains. Production for feasts (e.g., brewing, cooking, waste disposal) was thus more visible in the record of charred macroremains than actual rituals involving food and drink. The latter may require more and diverse lines of evidence to elucidate (e.g., residue study of drinking cups, serving vessels; extraction of microfossils from ritual contexts to identify food or drink offerings).

In his contribution, Morehart explores an important aspect of ritual practice that may be revealed through paleoethnobotany—the temporality of ritual. Arguing from the premise that ritual practices produce empirical data that may provide insight into both seasonality and symbolism, Morehart presents the case of an Epiclassic period shrine in the northern Basin of Mexico. The nondomestic nature of the deposits is clear: skull burials (with broken braziers and effigy figurines) are associated with freshwater springs. Pine charcoal and incense burner sherds were ubiquitous and associated: burning pine charcoal is inferred to be a fundamental ritual element. Maize and bean remains appear to be remains of food offerings. *Tagetes* (Aztec marigold) pollen is present in higher concentrations in levels associated with crania burials than in surrounding strata, providing convincing evidence that this flower was part of ritual. But while ecological seasonality indicators (maize and bean remains, marigold pollen) all suggest the *end* of the rainy season for the ritual activity, symbolism of the effigy figurines points to the *beginning* of the rains, prior to planting. The ambiguity created by these results—perhaps stored crops and dried flowers were used in a spring ritual—demonstrates the potential complexity and multi-layered meanings of ritual practices, and how these may be revealed through consideration of multiple lines of evidence.

Revealing the Social Realm Through Paleoethnobotany

Food is central to life, not just in staying alive, but in our relationships with other human beings. As Conkey emphasizes in her discussion of the intersectionality of paleoethnobotany, social theory, and feminist theory and practice, a focus on foodways provides archaeologists with an entry at the scale of the intimate, to social relationships and home life. She presents a compelling argument that a more social and "feminist-inspired" paleoethnobotany has expanded our knowledge of past human lives. Conkey illustrates this point in part by reviewing studies by Hastorf, Fritz, Morehart, and others, which demonstrate the deeper and more nuanced sense of people and plants that is gained when paleoethnobotanists are "thinking gender" and "thinking like a feminist".

In their overview of cultigen chenopods in the Americas, Fritz and colleagues compared and contrasted the sociocultural contexts in which these tiny seeds have been found in sites in North and South America, exploring the roles of chenopod in sociopolitical life in both regions. They provided a comprehensive review of morphological, molecular, and DNA approaches for distinguishing among species and varieties of chenopod, and then illustrated how these advances in identification clarified the roles of the crop in the food systems of eastern North America and the Bolivian *altiplano*. While there were similarities in the early history of chenopods in the two regions—initial appearances in disturbed habitats associated with early gardens (North America) or camelid corrals (Andes); contributor with other native plant and animal foods to communal gatherings and ceremonies of early complex societies—the trajectories of chenopod diverged later in time. Production of chenopod declined in North America several centuries after maize was intensified, while chenopod remained important after maize (especially maize beer) became central to life in the Andes, sustaining native peoples through the rise and fall of successive states and empires. Future research will hopefully explore further the ecological, agronomic, social, and political factors relevant to the changing role of chenopod in foodways and past human lives in these regions.

Each of the contributions to this volume illustrates how taking a viewpoint at the scale of the intimate gives new perspectives on the social realm. Through careful, systematic sampling for archaeobotanical remains in domestic and ritual spaces at Conchopata, Sayre and Whitehead were able to see the feast from the brewery: charred macroremains demonstrated the intimate relationships between domestic and ritual spaces, and by inference the activities carried out in them. Morehart's research at an Epiclassic period shrine in the northern Basin of Mexico illustrated how an action embedded in daily life—burning pine wood fuel—was transformed into a ritual act. In her research in northwestern Argentina, Korstanje viewed subsistence practices from the perspective of those laboring to collect and grow plants, investigating the effort, social networks, and negotiations put into procuring resources. The focus of Chevalier and colleagues was explicitly at the level of the intimate, to understand the lives of Middle Belgium's LBK peoples through multiple lines of botanical and other archaeological evidence. Similarly, Farahani and colleagues demonstrated what could be learned about culinary practices at La Joya de Ceren, El Salvador, through a focus on what could be reached from the metate.

Concluding Thoughts

In my commentary on the papers of this volume, I have focused on contemporary practices and approaches in social paleoethnobotany, or paleoethnobotany beyond diet, environment, and ecology. As I said at the outset, this was not my personal focus as a paleoethnobotanist, but I applaud and value these and other efforts to delve more deeply into the social worlds of past communities through archaeobotanical data.

Investigating the social world empirically through the remains of plants that played roles in daily life and ritual practices involves all the challenges of addressing diet, environment, and ecology through paleoethnobotany. For all of us, in the end it comes down to drawing convincing inferences about human behaviors from plants *present* at sites, and from the *combinations* in which they occur, and the *contexts* in which they are found. Fundamentally, this means inferring the processes and pathways through which the remains of plants—be they macroremains, pollen, phytoliths, starch grains, chemical signatures, and so on—became deposited and preserved in the sites and contexts we are studying.

Sometimes it is straightforward to model these processes and pathways. For instance, it is usually easy to distinguish modern and ancient plant tissues when only preservation of charred materials is expected. Microfossils embedded within burned vessel residues likely represent the remains of foods. Charred wood recovered from a hearth likely represents fuel. Comparing archaeological plant assemblages to modern surface vegetation provides insights for drawing inferences in more complex situations, for example, when ancient seeds are dried or waterlogged and not readily distinguishable from modern seeds. This approach also facilitates interpretation of pollen and phytolith data. Ethnographic observation, ethnoarchaeology, and experimental studies provide valuable insights into how plant tissues are transformed through cultural practices and potentially deposited and preserved. For example, it may be possible to infer the types of residues likely (and unlikely) to be recovered from a beer-brewing locality through study of traditional practices or the recreation of them. Scrutiny of the contexts in which remains occur (for example, inside versus outside structures) can also be critical in differentiating among depositional pathways. Micromorphological analysis, the study of diverse sediment, bioarchaeological, and microartifactual components in situ in large format thin sections, provides insights into formation processes of complex deposits.

Through these and related approaches, much has been learned about processes of deposition and preservation of macroremains, pollen, phytoliths, and starch, making it possible to model how these different kinds of plant remains are represented in the archaeological and geological records (Pearsall 2015). In general, the key factors for understanding representation of archaeobotanical remains differ by type: preservation and pollination biology for pollen, preservation and food processing technology for starch, production patterns in plants and tissues for phytoliths, and preservation and fire technology for macroremains. These understandings underpin inferences of human behavior drawn from the archaeobotanical record. For example, if our research question requires data on the social context and actions surrounding *consumption* of food or drink, understanding how food and drink are potentially represented as pollen, starch, phytoliths, or macroremains in such contexts is essential for identifying the actions. Drinking and pouring libations in a ritual space is unlikely to produce charred macroremains there, but might result in the deposition of phytoliths, starch, or pollen in the floor of the locality or in vessel residue. On the production side, tasks centered around a grinding stone might be more productively approached through a different set of indicators than those in taskscapes of beer brewing, tuber pit-roasting, or grain cleaning. While generalized models of representation may point us towards productive kinds of data to incorporate into our studies, I believe that the strongest inferences are those grounded in ethnographic observations, ethnoarchaeology, or experimental studies of human behaviors surrounding plants, and in a sound understanding of the archaeological contexts of samples.

I agree with Morehart and Morell-Hart's (2013) assessment that increased attention to method is critical if social paleoethnobotany is to address the kinds of questions central to social archaeology in a convincing way. They were speaking of research in Mesoamerica, but I think several points can be applied more generally. For example, more spatially systematic sampling within household contexts is needed to investigate patterns of food production and consumption, which are critical components for reconstructing social practices. Botanical data should be integrated with other archaeological indicators for better understanding of the functional nature of contexts; this understanding is essential for drawing robust inferences. Finally, as Christine Hastorf has long emphasized, identifying more diverse forms of plant remains is essential for "making the invisible visible" and placing food squarely in the center of social and political relations. Research that integrates multiple biological indicators provides the most promise for drawing strong inferences of past human behaviors. The more complex the behaviors, the greater the challenges, and the potential rewards.

References

Hastorf, C. A. (2001). Making the invisible visible: The hidden jewels of archaeology. In P. Drooker (Ed.), *Fleeting identities: Perishable material culture in archaeological research* (pp. 27–42, Center for Archaeological Investigations Occasional Paper 28). Carbondale, IL: Southern Illinois University Press.

Hastorf, C. A. (2003). Community with ancestors: Ceremonies and social memory in the Middle Formative at Chiripa, Bolivia. *Journal of Anthropological Archaeology, 22*(4), 305–332.

Hastorf, C. A. (2007). Archaeological Andean rituals: Performance, liturgy, and meaning. In E. Kryriakidis (Ed.), *The archaeology of ritual* (pp. 77–107). Los Angeles, CA: UCLA Cotsen Press.

Hastorf, C. A. (2008a). Food and feasting, social and political aspects. In D. M. Pearsall (Ed.), *Encyclopedia of archaeology* (pp. 1386–1395). Oxford: Academic Press.

Hastorf, C. A. (2008b). Paleoethnobotany. In D. M. Pearsall (Ed.), *Encyclopedia of archaeology* (pp. 1790–1795). Oxford: Academic Press.

Morehart, C. T., & Morell-Hart, S. (2013). Beyond the ecofact: Toward a social paleoethnobotany in Mesoamerica. *Journal of Anthropological Archaeology, 22*(2), 483–511.

Pearsall, D. M. (2015). *Paleoethnobotany: A handbook of procedures* (3rd ed.). Walnut Creek, CA: Left Coast Press.

[illegible]

[illegible]

[illegible]

[illegible]

[illegible]

[illegible]

[illegible] Oxford University Press.

Index

M.P. Sayre and M.C. Bruno (eds.), *Social Perspectives on Ancient Lives from Paleoethnobotanical Data*, DOI 10.1007/978-3-319-52849-6

The manufacturer's authorised representative in the EU is Springer Nature Customer Service Centre GmbH, Europaplatz 3, 69115 Heidelberg, Germany. If you have any concerns regarding our products, please contact ProductSafety@springernature.com

Printed and bound by CPI Group (UK) Ltd, Croydon, CR0 4YY
30/09/2026
02225643-0001